Bianca Adler

Boundary-Layer Processes Producing Mesoscale Water-Vapour Variability over a Mountainous Island

Wissenschaftliche Berichte des Instituts für Meteorologie und Klimaforschung des Karlsruher Instituts für Technologie (KIT)
Band 67

Herausgeber: Prof. Dr. Ch. Kottmeier

Institut für Meteorologie und Klimaforschung
am Karlsruher Institut für Technologie (KIT)
Kaiserstr. 12, 76128 Karlsruhe

Eine Übersicht aller bisher in dieser Schriftenreihe erschienenen Bände finden Sie am Ende des Buches.

Boundary-Layer Processes Producing Mesoscale Water-Vapour Variability over a Mountainous Island

by
Bianca Adler

KIT Scientific Publishing

Dissertation, Karlsruher Institut für Technologie (KIT)
Fakultät für Physik, 2014
Referenten: Prof. Dr. Ch. Kottmeier
PD Dr. M. Kunz

Impressum

KIT Scientific Publishing

Karlsruher Institut für Technologie (KIT)
KIT Scientific Publishing
Straße am Forum 2
D-76131 Karlsruhe

KIT Scientific Publishing is a registered trademark of Karlsruhe Institute of Technology. Reprint using the book cover is not allowed.

www.ksp.kit.edu

Print on Demand 2014

ISSN 0179-5619
ISBN 978-3-7315-0247-0
DOI: 10.5445/KSP/1000042282

Boundary-Layer Processes Producing Mesoscale Water-Vapour Variability over a Mountainous Island

Zur Erlangung des akademischen Grades eines

DOKTORS DER NATURWISSENSCHAFTEN

von der Fakultät für Physik des

Karlsruher Instituts für Technologie (KIT)

genehmigte

DISSERTATION

von

Dipl.-Met. Bianca Adler

aus Speyer

Tag der mündlichen Prüfung: 20. Juni 2014

Referent: Prof. Dr. Ch. Kottmeier

Korreferent: PD Dr. M. Kunz

Abstract

The water-vapour content in the lower and middle troposphere and atmospheric stratification are regarded as crucial for the development of isolated deep convection. Spatial inhomogeneities of the pre-convective atmospheric conditions over complex terrain occur under fair weather conditions, due to the evolution of the mountain Atmospheric Boundary Layer (mountain ABL). The mountain ABL results from the simultaneous occurrence of convection and mesoscale transport processes, which are associated with thermally driven circulations. However, the understanding of the superposition of these processes and of the impact of the mountain ABL on the evolution of deep convection is still limited, because observational data covering the different scales are rare. This especially applies to mountainous islands, which are known to be a preferred region for the evolution of deep convection. This thesis focused on the identification of processes relevant for the mountain ABL evolution over a mountainous island and on the evaluation of their impact on the spatial variability of water vapour, convection-related parameters and the evolution of deep convection by means of observations. Furthermore, the capabilities of combined measurement systems to capture multi-scale processes over complex terrain were assessed.

The investigation was based upon data analysis of observations obtained on the mountainous island of Corsica in the western Mediterranean Sea during the Hydrological cycle in the Mediterranean eXperiment (HyMeX) field campaign performed in late summer and autumn 2012. The Corsican Island features a high north-northwest to south-southeast oriented mountain ridge and the large Tavignano Valley in the northern part of the island. Thus, the island provides ideal conditions to address the above foci. The used data were mostly collected with the mobile observation platform KITcube, which combines various in-situ and remote sensing systems, like radiosonde systems, wind lidars, a microwave radiometer and a cloud radar. Measurements at two main deployment sites, one in the centre of the island in the Tavignano Valley and one on the east coast, were complemented by additional data from surface stations, a permanent Global Positioning System (GPS) network, which provided Integrated Water Vapour (IWV), and aircraft measurements. The designed measurement configuration and coordinated scan strategies successfully allowed to resolve relevant processes on different scales. Objective methods were developed and implemented to determine the height of the mountain ABL and convection layer, i.e. the part of the mountain ABL affected by surface-based, buoyancy-driven turbulent mixing.

Based on case studies and long-term observations experimental evidence of the characteristics of the mountain ABL and transport processes controlling its evolution was given. In the Tavignano Valley, undisturbed mountain ABL evolutions featured a convection layer being significantly lower than the mountain ABL, as topographic and advective venting decisively contributed to the evolution of a deep mountain ABL. Previously unreported processes were observed in the Tavignano Valley. These included strong subsidence in the order of -1 m s^{-1}, which developed within the mountain ABL in late morning and was induced by thermally driven low-level divergence, as well as vertical cell coupling between surface-based convective cells and elevated updraughts. The latter was associated with an increase of humidity in the layer with elevated updraughts which indicated an effective vertical transport of humidity. The evolution of the mountain ABL had a large impact on the humidity distribution over the island. On days dominated by the evolution of a mountain ABL, the IWV was characterized by a distinct diurnal evolution and spatial distribution. In areas over and downstream of the mountains which were affected by topographic and advective venting and by the evolution of a deep mountain ABL, the IWV strongly increased during the afternoon, reached maximum values in late afternoon and decreased afterwards. The mountain ABL also significantly impacted the pre-convective environment making it more favourable for the subsequent evolution of deep convection in the interior of the island than on the coasts. On some days, the intrusions of warm and dry air masses from the free atmosphere locally interrupted or inhibited the evolution of the mountain ABL on the downstream side of the main mountain ridge. These intrusions produced mesoscale variabilities in water vapour and atmospheric stratification, which were controlled from above, and caused locally unfavourable conditions for deep convection.

Contents

Abstract i

1 Motivation 1

2 Phenomenological and Theoretical Background 7

2.1 The Convective Boundary Layer over Horizontally Homogeneous and Flat Terrain 8

2.1.1 Structure 8

2.1.2 Diurnal Evolution 10

2.1.3 Height Determination 11

2.2 Thermally Driven Circulations 15

2.2.1 The Slope-Wind System 16

2.2.2 The Valley-Wind System 18

2.2.3 The Sea-Breeze Circulation 19

2.2.4 Superposition of Thermally Driven Circulations 20

2.3 Transport Processes over Complex Terrain and Definition of the Mountain ABL 22

2.4 Impact of the Background Flow on the Mountain ABL over Complex Terrain 26

2.4.1 Turbulent Transport 26

2.4.2 Dynamically Driven Flows 28

2.5 Atmospheric Water Vapour and its Detection . . . 36

2.6 Convection-Related Parameters 43

3 The HyMeX Field Campaign and Data Base 49

3.1 The HyMeX Field Campaign 49

3.2 Measurements on the Corsican Island during HyMeX 50

3.2.1 Terrain Characteristics and Station Locations 50

3.2.2 Instrument Specifications 52

4 Determination of the Convection-Layer and the Mountain ABL Height **67**

4.1 The Convection Layer 68
4.2 The Mountain ABL 76

5 The Evolution of the Mountain ABL **81**

5.1 The Undisturbed Mountain ABL Evolution: Case Study of 19 August 2012 82
5.1.1 Characteristics of the Mountain ABL at Corte 82
5.1.2 Processes in the Mountain ABL 91
5.1.3 Conceptual Model 107
5.2 The Interrupted Mountain ABL Evolution: Case Study of 02 October 2012 112
5.2.1 Large-Scale Conditions 112
5.2.2 Mountain ABL Evolution on the Coasts . . 115
5.2.3 Mountain ABL Evolution at Corte 117
5.2.4 Interpretation of the Observations 125
5.3 The Inhibited Mountain ABL Evolution: Case Study of 09 October 2012 128
5.3.1 Large-Scale Conditions 128
5.3.2 Mountain ABL Evolution over the Corsican Island 131
5.3.3 Interpretation of the Observations 137
5.4 Discussion 139

6 Effect of the Mountain ABL on Water Vapour and Atmospheric Stratification **145**

6.1 Variability of Water Vapour over the Corsican Island 145
6.1.1 Undisturbed Mountain ABL Evolution . . . 152
6.1.2 Isolated Deep Convection 165
6.1.3 Interrupted and Inhibited Mountain ABL Evolution 167
6.2 Convection-Related Parameters 169
6.2.1 Temporal Evolution at Corte 171
6.2.2 Spatial Distribution over the Corsican Island 174
6.3 Discussion 177

7 Summary and Conclusions **185**
7.1 Summary 185
7.2 Overall Conclusions 193

Acknowledgement **197**

A Acronyms and Symbols **199**

B Tables and Figures **205**

C Bibliography **223**

1. Motivation

Over complex terrain, deep convection and convective precipitation frequently occur in preferred areas. The western Mediterranean is known for both: during late summer and autumn, this area is often affected by heavy precipitation (Ducrocq et al., 2008), mainly convective in nature (Doswell III et al., 1998). Both the orography and the sea surface influence the formation and evolution of the convective systems in the region (Homar et al., 1999; Buzzi and Foschini, 2000; Rotunno and Ferretti, 2001). For example, the mountainous island of Corsica is regularly affected by heavy precipitation events (Lambert et al., 2011).

The development of deep convection requires some kind of atmospheric instability, a sufficient amount of humidity in the lower and middle troposphere and a trigger mechanism. In particular over mountainous islands, the regular formation and superposition of various thermally driven circulations cause convergence zones associated with upward motions and provide a trigger mechanism (Qian, 2008). In terms of the parcel theory, the trigger mechanism has to be strong enough to overcome Convective Inhibition (CIN) and to release its Convective Available Potential Energy (CAPE). Dry air in the middle troposphere may mix with the air in the rising parcel and may inhibit the evolution of deep convection, despite high CAPE and low CIN values (e.g. Adler et al., 2011b). To characterize the pre-convective atmospheric environment, various parameters are often analysed: CAPE and CIN describing conditional instability and the TT index accounting for static stability as well as humidity in the lower troposphere.

Those convection-related parameters were also calculated for the Corsican Island based on radiosoundings launched operationally by METEO FRANCE at Ajaccio on the island's west coast at around noon. As a measure of mid-level humidity the mean relative humidity (RH) between 850 and 700 hPa was additionally considered. The relation between these parameters and the occurrence of deep convection over the island for the months August to October was

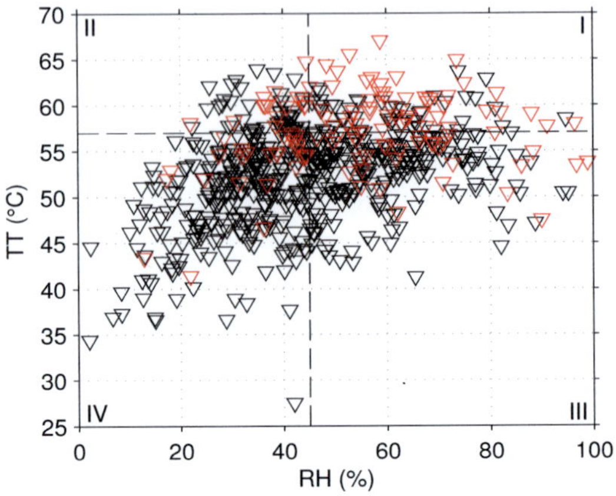

Figure 1.1.: Values of the TT index and RH at Ajaccio on days with (red) and without (black) isolated deep convection during the months August, September and October from 2001 and 2010. The parameters are derived from profiles measured by radiosondes. The dashed lines indicate the thresholds for the TT index and RH and the respective categories I to IV are indicated in the corners.

investigated for a 10-year time period from 2001 to 2010. As an indicator for deep convection, the cloud-to-ground lightning activity was used. The data derived from the Siemens lightning information service (BLIDS), which is based on the European Cooperation for Lightning Detection (EUCLID). Only lightning occurring after noon was considered, because locally triggered deep convection requires the earlier evolution of thermally driven circulations. Depending on whether lightning occurred and CAPE, CIN, TT index and RH were higher or lower than their respective thresholds, each day was classified. A CAPE of higher than 500 J kg^{-1}, a CIN of lower than 50 J kg^{-1}, a TT index of higher than 57 °C and a RH value of higher than 45 % were assumed to be favourable for deep convection. The statistical analysis led to some unexpected results: only about 16 % of all days with lightning occurred when CAPE, CIN and RH were classified as favourable for deep convection. All other days with lightning activity were characterized by at least one non-favourable parameter out of the three, e.g. about 20 % of all days with lightning were identified by low CAPE, high CIN and low RH values. The values of the TT index and RH were spread over a wide range, regardless of whether deep convection occurred on the respective day or not (Fig. 1.1). About 41 % of all days with lightning activity were associated with a high TT index and high RH (category I), i.e. conditions assumed to be favourable for deep convection. However, 13 % of the days when lightning was detected were characterized by a high TT index but low RH (category II), 25 % by a low TT index but high RH (category III) and 21 % by a low TT index and low RH (category IV). This means that no clear correlation between the occurrence of deep convection over the island and the atmospheric conditions on the west coast was found.

Parts of these unexpected results were possibly caused by frontal systems or large-scale advection, which modified the atmospheric environment after the radiosoundings were performed and made the conditions more favourable for deep convection. In addition to that, mesoscale spatial variabilities in the distributions of low-level humidity and temperature have a strong impact on the evolution of deep convection (e.g. Crook, 1996; Weckwerth, 2000), so that the radiosoundings on the west coast might not have been representative for the atmospheric conditions, in which the deep convection eventually evolved. In order to understand the mesoscale variability of atmospheric quantities, such as humidity and temperature, the processes affecting their temporal evolution and spatial distribution have to be considered.

Various processes ranging from the turbulent to the synoptic scale occur in the atmosphere and lead to a complex evolution and distribution of atmospheric quantities. Under fair weather conditions during daytime, which means that neither frontal systems nor large-scale advection significantly influence the temporal evolution of the local atmospheric conditions, the diurnal variation of atmospheric quantities over horizontally homogeneous and flat terrain is mainly determined by convection and restricted to the Atmospheric Boundary Layer (ABL). In this case the ABL, i.e. the layer directly affected by the presence of the Earth's surface, is also termed Convective Boundary Layer (CBL). Over complex terrain, thermally driven circulations are produced in addition to convection by differential heating of atmospheric layers. These circulations induce further mesoscale transport processes, such as topographic and advective venting. Due to the superposition of convection and mesoscale advection, the impact of the Earth's surface on the atmosphere over complex terrain extends over a deep atmospheric layer, named mountain ABL. The evolution and structure of the mountain ABL may significantly differ from the CBL over homogeneous terrain under equal surface forcing. The mountain ABL evolution may be additionally influenced by the background flow via turbulent transport or due to dynamically driven flows.

Although the existence of the various transport processes is known, the spatio-temporal variabilities of humidity and temperature and the impact of the transport processes on these variabilities are still not well understood. This is because complex terrain is often difficult to access and only few observational data have been collected so far, especially on the transition between turbulent and mesoscale processes. This raises the question of whether there are additional relevant processes occurring in the mountain ABL. It also needs to be investigated to what extent a mountain ABL affects the evolution of deep convection and if a detailed knowledge of the atmospheric conditions in the mountain ABL helps to understand the development of deep convection over complex terrain.

Because of these problems and open questions, the foci of this thesis are to identify multi-scale transport processes over a mountainous island and to assess their impact on the spatial variability of the convection-related parameters by means of comprehensive observations. The scientific objectives which are addressed in detail are:

1. To monitor the evolution of the mountain ABL under fair weather conditions and to determine the processes which are responsible for its evolution.

2. To analyse how the mountain ABL evolution is influenced by the background flow.

3. To investigate how the evolution of the mountain ABL affects the temporal and spatial variability of water vapour and atmospheric stratification over the Corsican Island and to study whether information on a higher spatial resolution allows a better understanding of the evolution of deep convection.

To approach these objectives on the basis of observations, simultaneous measurements of relevant processes on different scales are necessary. This must be realized by combining various instruments in such a way that they ideally complement each other. Hence, an additional focus of this thesis is to develop appropriate measurement configurations and scan strategies and to assess the capability of integrated measurement systems to capture processes in their entirety and to provide information required for the process understanding.

The various topics of this thesis were ideally addressed with measurements on the Corsican Island during the Hydrological cycle in the Mediterranean eXperiment (HyMeX; Ducrocq et al., 2014; Drobinski et al., 2014) field campaign in late summer and autumn 2012. The observations were mainly conducted with the mobile observation platform 'KITcube', which combines various in-situ and remote sensing instruments and has recently been developed to capture multi-scale atmospheric processes (Kalthoff et al., 2013a). The complex orography of the mountainous island of Corsica, the distribution of various measurement stations on the island and advanced measurement systems allowed to study processes and conditions in the mountain ABL by means of case studies as well as for the total duration of the field campaign.

In the following section, the phenomenological and theoretical background of the various aspects relevant for the subsequent data analysis is presented. In Sect. 3, the HyMeX field campaign and the used data are briefly overviewed. Objective methods to determine the structure of the mountain ABL are described in Sect. 4. The results from the data analysis are split into two sections: conditions and processes in the mountain ABL are analysed and discussed in consideration of different mountain ABL evolutions (Sect. 5). Section 6 addresses the impact of the mountain ABL on the variabilities of water vapour and convection-related parameters over the Corsican Island. Finally, the results are summarized and overall conclusions are given (Sect. 7).

2. Phenomenological and Theoretical Background

This thesis focuses on atmospheric processes over complex terrain under fair weather conditions during daytime ranging from the turbulent scale of a few metres to scales of several tenths of kilometres, i.e. from the micro-γ to the meso-γ scale (Orlanski, 1975). Several transport processes are investigated, including convection and advection. In meteorology, advection describes the transport of atmospheric variables, like humidity and heat, with the mean flow. Convection relates to the transport by the vertical flow component that results from buoyancy. It is distinguished between dry convection and moist convection. Dry convection means that there are no phase changes involved. It is generally highly turbulent and mainly occurs in the Convective Boundary Layer (CBL) (e.g. Emanuel, 1994). Moist convection is associated with the conversion of water vapour into liquid water, which is accompanied by a large amount of heat release. In the following, the term convection always refers to dry convection and stands for surface-based, buoyancy-driven turbulent mixing. When it is referred to deep moist convection it is explicitly stated.

In Sect. 2.1, a brief description of the structure, evolution and detection methods of the CBL over horizontally homogeneous and flat terrain is given, followed by an overview of various thermally driven circulations that evolve over complex terrain. In Sect. 2.3, the superpositions of convection and thermally driven circulations over complex are described and the mountain ABL is defined. Various ways how the background flow may impact the evolution of the mountain ABL over complex terrain are presented in Sect. 2.4. Finally, the atmospheric water vapour and its detection (Sect. 2.5) and parameters relevant for the evolution of deep moist convection (Sect. 2.6) are overviewed.

2.1. The Convective Boundary Layer over Horizontally Homogeneous and Flat Terrain

The Atmospheric Boundary Layer (ABL) may be defined as that part of the atmosphere which is directly affected by the presence of the Earth's surface and responds to surface forcing within a time scale of around an hour or even less (Stull, 1988). Consequently, diurnal variations of e.g. temperature and humidity close to the ground, which are not evident at higher levels, are one of the key characteristics of the ABL. Under fair weather conditions during daytime, the ABL is often termed CBL as its vertical structure is mainly determined by convection. Over horizontally homogeneous and flat terrain, the structure and processes in the CBL are mostly well understood.

2.1.1. Structure

Atmospheric turbulence may be produced by buoyancy, when thermals of warm air rise or sink because they are less or more dense than the ambient air, or by mechanical processes in layers with strong vertical wind shear. The vertical transport of atmospheric quantities in the CBL is mainly due to buoyancy-generated turbulent mixing. Within the 'classical' CBL three layers from bottom to top are distinguished (e.g. Stull, 1988): the surface layer, the well-mixed layer and the entrainment zone (Fig. 2.1). The surface layer is defined as the lowest layer of the CBL where the fluxes vary by less than 10 % with height. Typically, the surface layer composes 5 to 10 % of the CBL. Because the exchange of heat, humidity and momentum with the surface overcomes the effect of vertical mixing, it is characterized by an unstable stratification, humidity decrease with height and strong vertical wind shear. Under most conditions, the Monin-Obukhov similarity theory works well for the surface layer. The unstable conditions in the surface layer allow convection, which tends to produce roughly height-constant profiles of conserved quantities, like potential temperature and specific humidity, in the well-mixed or mixed layer. Also aerosols are often well mixed throughout that layer, whose top is designated z_1. Thermals from the surface layer are positively buoyant

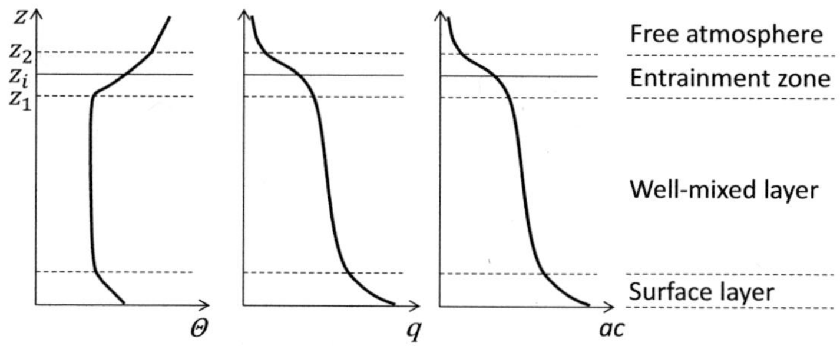

Figure 2.1.: Daytime profiles of mean potential temperature, Θ, specific humidity, q and aerosol content, ac, in a typical CBL over horizontally homogeneous and flat terrain. The characteristic layers and the height of the well-mixed layer, z_1, the CBL height, z_i and the height where the free atmosphere starts, z_2 are indicated (adapted from Stull, 1988).

when they rise through the well-mixed layer. Once the thermals reach the warmer free atmosphere, they become negatively buoyant, but overshoot their equilibrium level because of their vertical momentum. As there is generally little turbulence outside the well-mixed layer, the thermals do not mix with the ambient air and sink back into the well-mixed layer. During the overshooting potentially warm, dry and aerosol-free air from the free atmosphere is transported into the well-mixed layer and becomes rapidly mixed down. Typically, it takes about 10 to 20 minutes for air to circulate between the surface and the top of the well-mixed layer. Because of entrainment as well as evapotranspiration and aerosol emissions at the surface, specific humidity and aerosol content often slightly decrease with height in the well-mixed layer. The well-mixed layer is topped by a stably stratified layer, the entrainment zone or interfacial layer, which is characterized by mixing of free atmosphere air downwards and by overshooting thermals. The depth of the entrainment zone varies a lot and can take 10 to 60 % of the CBL. Its top is named z_2.

Multiple definitions of the CBL height, z_i, exist in literature, for which it is claimed that $z_1 \leq z_i \leq z_2$. For example, z_i is defined as the level of most negative sensible heat flux (e.g. Stull, 1988); as the bottom of the lowest inversion (e.g. Kalthoff et al., 1998); the level at which the potential temperature of a rising air parcel equals the ambient potential temperature (e.g. Hennemuth and Lammert, 2006); or the middle of the entrainment layer, i.e. $z_i = \frac{1}{2}(z_1 + z_2)$ (e.g. Steyn et al., 1999). Different methods to detect z_i are presented in Sect. 2.1.3.

2.1.2. Diurnal Evolution

Under fair weather conditions the ABL over horizontally homogeneous and flat terrain shows a strong diurnal cycle. The typical diurnal development is well described in textbooks like Stull (1988) and Garratt (1994). The breakup of the nocturnal surface inversion (A in Fig. 2.2) starts shortly after sunrise, when an upward surface sensible heat flux warms the air near the surface. A CBL with a surface layer and a well-mixed layer starts to develop, being

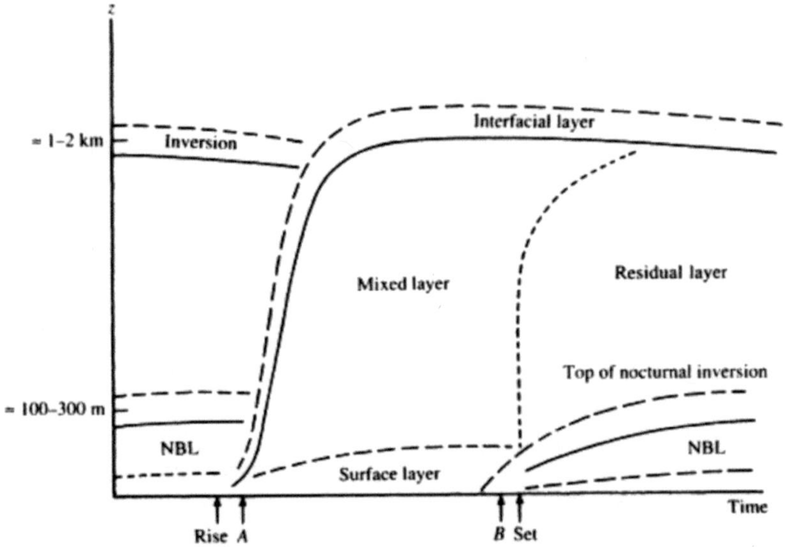

Figure 2.2.: Schematic diagram of the diurnal ABL evolution over horizontally homogeneous and flat terrain under fair weather conditions (Garratt, 1994). 'Rise' indicates the sunrise, 'Set' the sunset, 'A' the beginning of the breakup of the nocturnal surface inversion and 'B' the time, when turbulence starts to decay. NBL denotes the nocturnal boundary layer.

shallow at the beginning. Entrainment of warmer air from above leads to a growth of the well-mixed or mixed layer. At first, it deepens slowly because it grows into the stably stratified nocturnal boundary layer. After the breakup of the nocturnal surface inversion, the well-mixed layer reaches the bottom of the residual layer. Subsequently, no inversion tops the well-mixed layer and thermals penetrate quickly upward, which leads to a rapid growth of the well-mixed layer. Once it extends to the capping inversion at top of the residual layer, the well-mixed layer growth rapidly slows down, leading to rather constant CBL heights for most of the afternoon. Shortly before sunset (B in Fig. 2.2), turbulence decays and the well-mixed layer is reclassified as residual layer, because the mean atmospheric variables remain more or less the same. With the cooling of the lower part of the residual layer due to a downward surface sensible heat flux, a stably stratified nocturnal boundary layer develops during the night.

2.1.3. Height Determination

Various definitions of z_i and methods for its detection have been developed in the past, which are in the strict sense only valid for horizontally homogeneous and flat terrain. Detailed overviews are given by e.g. Beyrich (1997) and Seibert et al. (2000). The definition of z_i often depends on the used data base and the heights derived from measurements based on different systems may vary a lot. Beyrich (1997) suggests two main reasons: different instruments measure different atmospheric variables with a different height resolution and measurement accuracy. Secondly, the structure of the CBL depends on multiple processes, e.g. turbulence, radiation, advection, vertical motions, etc., and these processes may impact the profiles of different mean and turbulent variables in various ways. Available measurement platforms include radiosonde, tethersonde and aircraft, which in-situ measure atmospheric parameters, or remote sensing systems, like lidar, sodar and wind profiler. The criteria to determine z_i may be summarized into two groups (Beyrich, 1997): the first one is based on profiles of mean variables, e.g. temperature, humidity or aerosol content. In doing so, it is assumed that convection dominates and these variables reflect the result of vertical mixing. The second one is based on profiles of turbulent variables, e.g. fluxes or the variance of the vertical wind speed, under the assumption that turbulence leads to mixing of atmospheric quantities. Standard methods to determine z_i based on radiosonde and lidar measurements are described in the following. In Sect. 4.1 some of these methods or modifications of them are applied to the used data set.

Mean Variables measured by Radiosonde

To determine z_i from mean profiles of temperature, humidity and wind derived from radiosoundings various methods have been developed (e.g. Seibert et al., 2000). For example, Rampanelli and Zardi (2004) suggest the use of best-fit analysis of soundings with a smooth ideal profile, which consists of a well-mixed layer with height-constant potential temperature, a stably stratified entrainment layer and a free atmosphere with a constant lapse rate. Other objective methods include e.g. the parcel method, the Richardson number method or the gradient method for potential temperature or specific humidity (Beyrich and Leps, 2012). The parcel method is based on the assumption, that an air parcel from near the surface will rise as long as it is warmer than the surrounding air. The layer where the air parcel's potential temperature agrees with the environmental potential temperature is then defined as z_i. Beyrich and Leps (2012) calculate the bulk Richardson number

$$Ri_b(z) = \frac{gz}{\overline{\Theta}_0} \frac{\left(\overline{\Theta}(z) - \overline{\Theta}_0\right)}{\overline{u}(z)^2 + \overline{v}(z)^2}, \tag{2.1}$$

with the gravity acceleration, g, the height, z, the near-surface mean potential temperature, $\overline{\Theta}_0$, the mean potential temperature profile, $\overline{\Theta}(z)$ and the profiles of the mean horizontal wind components, $\overline{u}(z)$ and $\overline{v}(z)$. These authors define z_i as the height where Ri_b exceeds the critical value of 0.2. Thus, the parcel method represents a special case of the Richardson number method with a critical value of zero. A disadvantage of these methods is the use of the near-surface temperature, which depends a lot on the local conditions, is sensitive to surface inhomogeneities and is often erroneous, due to insufficient ventilation of the sonde. Two other methods to determine z_i are based on the detection of the level with the maximum temperature or humidity gradient. Beyrich and Leps (2012) suggest an algorithm to detect these z_i, which includes the parcel method and the Richardson number method.

At first, the authors detect z_i with the parcel method and the Richardson number method and then they calculate the arithmetic mean between these two heights. This mean height plus 200 m is then used as an upper limit, below which the heights of the maximum gradients of potential temperature or specific humidity are calculated to derive two additional z_i. If neither the parcel method nor the Richardson number method provide a z_i, an upper limit of 3000 m Above Ground Level (AGL) is taken. This shall ensure that no elevated strong inversions are misleadingly taken as z_i.

Aerosol Content measured by Wind Lidar

As the Earth's surface is assumed to be the main source for aerosols and the temperature inversion at z_i represents a capping layer, the aerosol content below z_i is typically higher than in the layers above. Because of that, z_i is often associated with strong gradients of aerosol content and profiles of aerosol content measured by lidars are used to determine z_i. To detect the top of the aerosol layer various methods exist, which are described e.g. by Emeis et al. (2008) and Träumner et al. (2011). The methods include the threshold method, which is based on the definition of a threshold of the backscatter intensity; the gradient method, which defines the largest negative peak of the first derivative of the backscatter intensity as z_i; the idealized profile method, where z_i is determined from an idealized backscatter profile, which is fitted to the observed profile; and the wavelet method, which is based on a wavelet analysis of the backscatter intensity.

Variance of Vertical Wind Speed measured by Wind Lidar

Another approach to determine z_i is based on profiles of turbulent variables. When operated in vertical stare mode, remote sensing instruments, like wind lidars, provide the variance of the vertical wind speed

$$\sigma_w^2 = \overline{w'w'}, \tag{2.2}$$

where w' is the vertical wind speed fluctuation and the overbar indicates a time average. This variable is a measure of turbulence intensity according to the Turbulent Kinetic Energy (TKE) per unit mass M (e.g. Stull, 1988)

$$\frac{\text{TKE}}{M} = \frac{1}{2}\left(\sigma_u^2 + \sigma_v^2 + \sigma_w^2\right) \tag{2.3}$$

with σ_u^2, σ_v^2 being the variances of the horizontal wind components.

Using σ_w^2 profiles measured by wind lidars, z_i is determined under the assumption that no significant turbulence occurs above z_i. In the ideal case, σ_w^2 follows an idealized profile in the CBL (Lenschow et al., 1980)

$$\frac{\sigma_w^2}{w_*^2} = 1.8\left(\frac{z}{z_i}\right)^{\frac{2}{3}}\left(1 - 0.8\frac{z}{z_i}\right)^2, \tag{2.4}$$

with the convective velocity, w_*. According to the idealized σ_w^2 profile, the maximum σ_w^2 occurs in a height $d\,z_i$ with typical values for d between 0.35 and 0.4. Different approaches to determine z_i are outlined by e.g. Träumner et al. (2011). They are based on (i) an empiric threshold $\sigma_{w,\text{th}}^2$, (ii) the height of the maximum σ_w^2 or (iii) fitting an idealized profile of σ_w^2 to the measured one. In approach (i), z_i is determined as the height where σ_w^2 falls below an empiric threshold $\sigma_{w,\text{th}}^2$. This threshold depends on the atmospheric conditions and ranges from $\sigma_{w,\text{th}}^2 = 0.04$ m^2 s^{-1} for a maritime boundary layer (Tucker et al., 2009) and $\sigma_{w,\text{th}}^2 = 0.16$ m^2 s^{-1} for a CBL (Träumner, 2012). In approach (ii), the height of the maximum σ_w^2 is determined and z_i is then extrapolated from this height. When fitting the idealized profile of σ_w^2 (Eq. 2.4) to the measured one, z_i results as one of the fitting parameters in approach (iii). Träumner et al. (2011) found that approaches (ii) and (iii) strongly depend on the σ_w^2 profile in the lower part of the CBL. This is especially problematic for low z_i, because the lowest measurement range gate of the wind lidar is often around 400 m AGL. Therefore, approach (i) was used in thesis.

Advantages and Shortcomings of the Different Methods

All the above described methods using different devices and parameters have advantages and shortcomings, which are overviewed by e.g. Beyrich (1994) and Seibert et al. (2000). Only the most relevant for this thesis are listed below: a great advantage of using radiosonde profiles is the independency of aerosol content and a measurement range throughout the free atmosphere. However, they only provide a 'snapshot'-like profile of the atmospheric structure and the frequency of radiosoundings is limited, even during field campaigns, when they are often launched every couple of hours. Depending on the characteristics of the thermal the radiosonde flies through, the measured z_i may vary a lot temporally as well as spatially. Because of this, Stull (1988) even recommends not to use a single radiosounding to estimate z_i.

Aerosol content profiles measured by lidar provide z_i with a very high temporal resolution and therefore allow a very detailed detection of the CBL evolution. However, the lowest measurement range gate of a lidar prevents the detection of low z_i and residual layers with high aerosol content may lead to erroneous z_i (Träumner et al., 2011).

Using profiles of mean variables and aerosol content works well under certain meteorological circumstances. However, in cases with weak gradients at the CBL top, with non-perfectly mixed CBLs or when advection modifies the vertical distribution of potential temperature, specific humidity and aerosol content, these methods may be problematic and lead to significant differences in z_i (Seibert et al., 2000; Kalthoff et al., 2013b). Using vertical wind speed measured by lidar avoids these problems, as a lidar directly measures a turbulent variable and therefore detects turbulent vertical mixing. However, σ_w^2 is calculated for time intervals, which leads to a lower temporal resolution for z_i. If these time intervals are chosen too long, diurnal changes in turbulence during the day may not be resolved (e.g. Träumner, 2012). On the other hand, if they are chosen too short, large eddies may not be included in σ_w^2. Like for the aerosol content, the lowest range gate of a wind lidar limits the detection of low z_i.

2.2. Thermally Driven Circulations

Over complex terrain, thermally driven circulations on different scales develop under fair weather conditions. They are characterized by a reversal of wind direction twice per day (Whiteman, 2000). Differential heating of atmospheric layers during daytime and cooling during nighttime produces horizontal temperature gradients. These result in horizontal pressure gradients, which cause air close to the surface to flow from areas with lower temperatures and higher pressure to areas with higher temperatures and lower pressure. This flow represents the lower branch of a closed circulation. Ideally, the upper branch of this circulation, often called return flow, runs in the opposite direction from the flow in the lower branch. Large variations in speed, depth, onset time and duration of thermally driven circulations can occur from place to place, since they strongly depend on factors like terrain characteristics and the surface energy balance. Typical thermally driven circulations are slope-wind systems, valley-wind systems and sea-breeze circulations.

2.2.1. The Slope-Wind System

The lower branch of the slope-wind system is represented by an upslope wind during daytime, which blows up the slope, and by a downslope wind during nighttime, which blows down the slope (e.g. Whiteman, 2000). During the night, air near a slope is radiatively cooled and becomes negatively buoyant. During daytime, a temperature difference develops between the radiatively heated air in the boundary layer near a slope and the air at the same level over the valley. The driving force for upslope winds is the buoyancy force with an additional along-slope pressure gradient arising when the temperature perturbation increases going up the slope (Haiden, 2003). Upslope winds cannot approach a steady state, but grow continuously in strength and time as long as the surface sensible heat flux increases with time (e.g. Zardi and Whiteman, 2013). Over a slope, a convective-advective boundary layer develops, with bent-over plumes and convective cells moving up the slope. Upslope winds occur on a small scale and react instantly on short-term variations in the surface energy balance (Whiteman, 2000). Small-scale variability in surface conditions can produce local variations in direction and strength of the upslope winds, and even reverse their sign. Upslope winds typically reach maximum wind speeds of 1 to 5 m s^{-1} in about 10 to 50 m above ground, while the total depth of the upslope wind layer ranges from 20 to 200 m (Zardi and Whiteman, 2013). They often increase in depth as well as in speed with distance up the slopes. The strongest upslope wind speeds are typically reached in mid-morning when the horizontal temperature difference between the heated slopes and the ambient atmosphere is strongest. The large depth of the upslope-wind layer, its non-stationarity and its inhomogeneity make continuous observations of the whole upslope wind layer difficult and the process understanding still needs improvement. Return flows of several hundred metres of depth and wind speeds of 1 to 2 m s^{-1} were observed with theodolite measurements (Davidson, 1963; Wooldridge and McIntyre, 1986). More recently, return flows in upslope wind systems were described by e.g. Reuten et al. (2005), De Wekker (2008) and Serafin and Zardi (2010a).

In a valley, upslope winds are associated with divergence close to the valley bottom. This results in compensating subsidence in the valley core, which has already been postulated by Defant (1949). Observed warming in the valley core has been interpreted as an indirect measure of the subsidence (Whiteman, 1982; Vergeiner and Dreiseitl, 1987; Kondo et al., 1989). During a dry adiabatic descent, air maintains its potential temperature and thus potentially warmer air is transported towards the ground in a stably stratified atmosphere. In numerical simulations, the subsidence occurs as part of the closed slope wind circulation (Rampanelli et al., 2004; Serafin and Zardi, 2010a). The effect of the slope-wind system on the atmospheric stratification within a valley is twofold (Whiteman, 1982; Whiteman and McKee, 1982): compensating subsidence contributes to the breakup of the nocturnal surface inversion, but also reduces the growth of the CBL in the valley in the morning. The breakup of the nocturnal surface inversion requires the warming of the air in the valley until a height-constant layer forms in which the potential temperature has the same value as the inversion top at sunrise. In a valley, the growth of the CBL and the descent of the inversion top caused by the compensating subsidence both contribute to the inversion breakup in the morning. Using idealized large-eddy simulations, Serafin and Zardi (2010a) examined the heat transfer from close to the ground surface to the core of the valley atmosphere and identified two warming mechanisms, subsidence and convection. The CBL growth in the valley is reduced, as some of the available energy drives mass up the slopes and is thus not available for the CBL growth (Whiteman, 1982). Kondo et al. (1989) observed that subsidence in a basin depresses the evolution of a CBL and contributes to the warming of the basin atmosphere.

2.2.2. The Valley-Wind System

During daytime, a valley-wind system develops when the air in a valley is warmer than the air further down the valley or outside the valley over the adjacent plain (e.g. Zardi and Whiteman, 2013). The upvalley wind represents the lower branch of this circulation. During nighttime, when the temperature gradient is reversed, a downvalley flow evolves. An upvalley wind is driven by a hydrostatically produced horizontal pressure gradient, since the air in the valley warms more rapidly than the air over the plain. The differential warming may be explained by several mechanisms: in case of weak background winds, the surrounding orography protects the valley atmosphere from mass and heat exchange with the free atmosphere, which concentrates the heating in the valley during daytime. Because slopes partially enclose the valley air mass, radiative processes are more effective within the valley than over the plain. Presumably more important for the enhanced warming in the valley is another mechanism, named 'Topographic Amplification Factor' (TAF), described in detail by e.g. Whiteman (2000). This concept was first explored by Wagner (1932) and was further investigated by Steinacker (1984). The warming produced by incoming solar radiation depends on the air volume to be warmed. A volume of air having the same depth and area at its top is larger over a plain compared to a valley, because of the confining slopes. Thus, the air in the valley warms more during the day. However, when applying this concept in practice various difficulties arise, including the definition of the volume top, the volume of tributary valleys (Steinacker, 1984) and heat exchange through the volume top due to upslope flows or compensating subsidence (e.g. Rampanelli et al., 2004; Weigel et al., 2007b; Serafin and Zardi, 2011). The onset of the upvalley wind typically occurs several hours after the upslope winds are initiated (e.g. Zardi and Whiteman, 2013), i.e. their response time to changes in the surface energy balance is larger than for upslope winds. Typical upvalley wind speeds range from 3 to 10 m s^{-1}. Since the upper branch of the closed valley-wind circulation is mostly unconfined by orography, it is in general broader in its horizontal extent, weaker than the lower branch and often obscured by large-scale winds. Thus, it is reported infrequently. However, some observations of return flows exist and wind speeds of 1 to 2 m^{-1} were reported (Buettner and Thyer, 1965; Reiter et al., 1984; McGowan, 2004). The structure of the upvalley-wind system is additionally complicated by tributary valleys, as the upvalley flow divides at each juncture of a side valley with the main valley (e.g. Zardi and Whiteman, 2013). The mass flux in the main valley is reduced at these junctures due to upvalley flow in the tributary valleys (Zängl et al., 2004). Freytag (1987) reported how subsidence over the main valley compensates for upvalley wind in a tributary valley.

2.2.3. The Sea-Breeze Circulation

A sea-breeze circulation results from the horizontal pressure difference that develops between land and an adjacent sea during the day. Because water has a higher specific heat capacity and a higher heat conductivity than soil, a unit mass of water will warm less than a unit mass of soil, when they receive the same heat input (e.g. Whiteman, 2000). This leads to a rapid increase of the air temperature over a land surface during the day and a strong diurnal temperature variation, while the air temperature over the sea changes only little. Strong temperature and pressure gradients close to the surface develop across a coastline with lower pressure over the warmer land surface during daytime, causing the sea breeze to flow from the sea towards the land. During nighttime, the temperature and pressure gradients are reversed and a land breeze develops, which is usually weaker than the sea breeze, because of a weaker horizontal temperature gradient. The horizontal temperature gradient and therefore the strength and duration of a sea breeze depend on the season and the time of the day. In midlatitudes, sea breezes are typically strongest in spring and summer. The summer sea breeze is usually established by midmorning and the maximum wind speeds are reached in mid-afternoon. Maximum wind speeds of 2 to 10 m s^{-1} were reported in sea breezes all over the world (Atkinson, 1981, p. 145). The landward edge of the sea breeze, often called sea-breeze front, is characterized by sharp changes in temperature, humidity and wind, because of the cooler and moister air in the sea breeze that originated over the water. With the arrival of the sea-breeze front, the wind turns to onshore direction and typically accelerates. The propagation speed of the sea-breeze front, u_{sbf}, i.e. the rate at which it advances inland, can be related to the wind speed in the sea breeze, u_{sb} (e.g. Simpson and Britter, 1980). Based on approaches of Keulegan (1957) and Simpson (1969), Simpson and Britter (1980) determined the propagation speed of the sea-breeze front to

$$u_{sbf} \approx 0.87 u_{sb} + 0.59 u_g, \quad (2.5)$$

taking into account the cross-shore wind component, u_g, resulting from the large-scale pressure gradient. The two constants in Eq. 2.5 were gained in laboratory studies. Finkele (1998) used aircraft data to estimate the propagation speed from the onshore extent of the sea breeze and found typical values of 1 to 2 m s^{-1}, slowing down in the afternoon. Other studies reported propagation speeds ranging from 3 m s^{-1} in England (Simpson et al., 1977) to 7 m s^{-1} in Australia (Clarke, 1955). Inland penetration depths of about 150 km were observed in Spain (Kottmeier et al., 2000). They vary from 15 to 200 km as listed by Miller et al. (2003). Typical sea-breeze depths range from 100 to 1000 m (Atkinson, 1981, p. 143). Above the sea breeze a compensatory return flow exists as the upper branch of the closed sea breeze circulation. Although this flow is not always observed, return wind speeds ranging mainly from 2 to 4 m s^{-1} were reported (Atkinson, 1981, p. 145).

2.2.4. Superposition of Thermally Driven Circulations

Slope and Valley Winds

The slope and valley winds described in Sects. 2.2.1 and 2.2.2 do not occur separately, but superimpose and produce a coupled slope- and valley-wind system, which was first explored by Defant (1949). During a typical day in an idealized valley, four distinct phases occur (e.g. Zardi and Whiteman, 2013):

Morning transition phase Shortly after sunrise, an upward surface sensible heat flux transfers heat from the surface to the air above, leading to the evolution of a warm boundary layer. This marks the beginning of the morning transition phase. Upslope winds start to develop. The breakup of the nocturnal surface inversion and the warming of the valley atmosphere is mainly a result of two processes: the CBL growth in the valley and compensating subsidence (Sect. 2.2.1). The warming is amplified due to the valley orography (TAF concept, Sect. 2.2.2). Eventually, upvalley winds evolve. The complete breakup of the nocturnal surface inversion marks the end of the morning transition phase.

Daytime phase The daytime phase typically lasts from mid- to late morning until the late afternoon or early evening, when the outgoing longwave radiation exceeds the incoming shortwave radiation. During this period, the convection from the heated surfaces grows and a vertical coupling of the wind in the valley to the background flow is possible. Depending on the wind direction and speed of the background flow, this may lead to a turning of the wind and/or to a wind speed increase in the valley in the afternoon (Whiteman, 2000). The horizontal temperature gradients along the valley axis and thus the upvalley wind are maintained, which may be explained in terms of the TAF concept and radiative processes (e.g. Zardi and Whiteman, 2013).

Evening transition phase Starting in areas of the valley, which are shadowed by terrain, a downward surface sensible heat flux cools the air above the surface. Downslope winds start and compensating rising in the valley centre and cold air advection with the still existing upvalley wind cool the valley atmosphere. Typically, the upvalley wind prevails for several hours after a downward surface sensible heat flux developed. Eventually the pressure gradient between the valley and adjacent plain reverses and a downvalley flow evolves. With the stabilization of the valley atmosphere during the evening transition period, the downslope winds become weaker and shallower.

Nighttime phase The nighttime phase is mainly characterized by the predominance of a downvalley wind. The phase typically starts sometime between several hours after sunset and midnight and lasts until the morning transition phase.

Interactions over Coastal Mountains

Over coastal mountains, additional interactions between the sea breeze and the upvalley and upslope winds occur. Coastal valleys funnel sea breezes and accelerate the flow up the valley (Miller et al., 2003). This results in a deeper inland penetration of the sea-breeze front than it would be the case over horizontally homogeneous terrain. Numerical simulations indicated that the interaction between an upvalley wind and a sea breeze results in an enhanced flow in the valley ahead of the sea-breeze front (Kondo, 1990a,b). Physick and Abbs (1992) reported sea breezes, which regularly penetrate more than 100 km up the valley. Khodayar et al. (2008) observed a superposition of an upvalley wind and sea breeze, which produces an effective inland transport of moist air. As the maritime air mass is often stably stratified (e.g. Bischoff-Gauß et al., 2006), the superposition of the upvalley wind and sea breeze leads to horizontal advection of a cool, stably stratified layer, which limits the depth of the well-mixed layer in the valley (Cox, 2006; Bergström and Juuso, 2006; Bischoff-Gauß et al., 2008; Khodayar et al., 2008) in addition to other processes related to compensating subsidence (Sect. 2.2.1).

Coastal slopes may either enhance or suppress sea breezes. When the temperature variation of the slope and the coastal plain are in phase, the sea breeze is amplified and occurs earlier (Miller et al., 2003). Banta et al. (1993) observed a faster growth of the sea breeze in the vicinity of a slope using a scanning wind lidar. On the other hand, the slopes may mechanically block the sea breeze. In an overview given by Crosman and Horel (2010) crucial factors determining the dependency of the sea breeze on terrain are described as e.g. the terrain slope, the terrain height or the location of the slope relative to the coastline.

2.3. Transport Processes over Complex Terrain and Definition of the Mountain ABL

The transport of air masses under fair weather conditions during daytime occurs on various scales, ranging from the turbulent scale to the mesoscale. Over horizontally homogeneous and flat terrain,

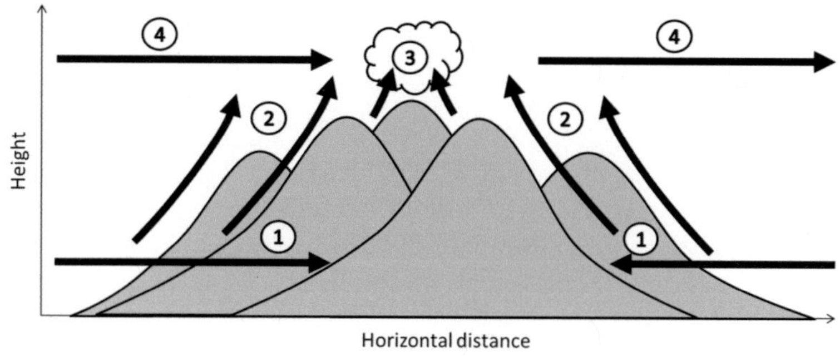

Figure 2.3.: Schematic diagram of transport processes occurring over complex terrain. The depicted processes are: horizontal transport with thermally driven winds (1), topographic venting (2), moist convection (3) and advective venting (4).

convection dominates the vertical transport, diurnal variations of atmospheric variables and the evolution of a CBL. Over complex terrain, mesoscale transport processes associated with thermally driven circulations additionally exist and superimpose convection. Most of the relevant transport processes are overviewed in Fig. 2.3. In particular over coastal mountains, sea breezes and upvalley winds often superimpose and provide for an effective horizontal inland transport of moist air (Physick and Abbs, 1992; Khodayar et al., 2008). Thermally driven winds transport air close to the surface upwards over the mountain ridges (e.g. Fast and Zhong, 1998), hereafter referred to as 'topographic venting'. Convergence of the thermally driven winds (e.g. Lu and Turco, 1994) or the formation of moist convection (e.g. Ching et al., 1988; Kalthoff et al., 2013b) above the ridges enhance the vertical transport. Over complex terrain, horizontal gradients of humidity, heat or aerosols develop due to an inhomogeneous CBL height or topographic venting, and horizontal advection with the mean wind occurs (e.g. Kossmann et al., 1999). The latter process is often referred to as 'advective venting'. The mean wind can either be a result of mesoscale processes or the large-scale conditions. As part of a closed thermally driven circulation, subsidence in a valley core compensates flow divergence close to the surface (e.g. Whiteman, 1982) and advects potentially warmer air towards the valley floor in a stably stratified atmosphere (e.g. Rampanelli et al., 2004; Serafin and Zardi, 2010a).

The impact of the different transport processes on the atmospheric structure over complex terrain was investigated in numerous observational and numerical studies. Airborne lidar observations revealed layers with a high aerosol content up to 4000 m above Mean Sea Level (MSL) over the Alps, i.e. more than 1000 m above ridge height, which extended well above the CBL (De Wekker et al., 2004; Henne et al., 2004). These authors attributed the upward transport of aerosols to thermally driven circulations, initiated by orography. The transport and diffusion of air masses with high ozone and pollutant concentrations due to topographic venting were observed for low mountain ranges (e.g. Fiedler et al., 2000; Kalthoff et al., 2000) and high mountain ranges, like the Alps (e.g. Furger et al., 2000; Prévôt et al., 2000). Near coastal mountains, interactions of sea breezes and slope winds lead to the formation of elevated pollution layers (e.g. Lu and Turco, 1994; Bischoff-Gauß et al., 1998). Once the polluted air masses intrude synoptically influenced layers, they get downstream advected with the mean wind forming elevated layers (e.g. Millán et al., 2002; Nyeki et al., 2002; Henne et al., 2004; Chen et al., 2009). When the stratification in elevated boundary layers is maintained during the horizontal transport, elevated mixed layers form downstream of the mountain ridge (e.g. Arritt et al., 1992; Sun and De Wekker, 2011).

The transport of humidity is significantly enhanced over complex terrain compared to horizontally homogeneous and flat terrain. Weigel et al. (2007a) found that the contributions to the humidity transport due to a narrowing of the valley cross-section and from thermally driven circulations exceed the contribution of turbulent mixing by several times. Upslope winds transport humidity from the valley over the ridges, reducing the Integrated Water Vapour (IWV) in the valley and increasing it over the mountains (Kuwagata et al., 2001). In case the evapotranspiration from the Earth's surface is lower than the humidity removal, Takagi et al. (2000) observed that the diurnal variation of IWV in the valley has its minimum in late afternoon. Over the mountains, IWV reaches its maximum in late afternoon (Wu et al., 2003). Extended sea breezes over a mountainous island enhance this IWV increase (Ohtani, 2001). Near the coast, sea breezes transport moist air from the sea and contribute to an IWV increase (Bastin et al., 2007). Topographic and advective venting result in the formation of elevated humidity layers downstream of the mountain ridge (e.g. Ohtani, 2001; Henne et al., 2005; Sasaki et al., 2004) and produce an abrupt increase of IWV over a semi-basin from evening to midnight (Iwasaki and Miki, 2001).

Large differences between the evolution of a CBL over complex and horizontally homogeneous terrain were found. Low (e.g. Rampanelli et al., 2004; Rampanelli and Zardi, 2004; Rotach and Zardi, 2007; Bischoff-Gauß et al., 2008) or non-existent (e.g. Khodayar et al., 2008) well-mixed layers were reported in valleys. A transient depression of the growth of the well-mixed layer occurs at the foot of isolated slopes due to vertical and horizontal advection of heat (De Wekker, 2008; Serafin and Zardi, 2010b). Above the well-mixed layer, a slightly stably stratified layer topped by an inversion and a more stably stratified free atmosphere were identified (e.g. De Wekker et al., 2004; Henne et al., 2004; Serafin and Zardi, 2010a). Weigel and Rotach (2004) and Weigel et al. (2006) reported a stabilization of the valley atmosphere. In this case, the stabilization is caused by vertical warm-air advection as part of a curvature-induced secondary circulation. Due to the centrifugal force cross-valley density differences evolve, causing a hydrostatic pressure gradient force. Because the centrifugal force increases with height, while the pressure gradient force is height-independent, a secondary circulation develops. In the same studies, significant turbulence in aircraft measurements was observed, despite the slightly stable stratification, and the authors proposed that profiles of potential temperature may not be a good indicator of the CBL height over complex terrain.

Thus, there is evidence that over complex terrain (i) the CBL structure and development differs from the classical image found over horizontally homogeneous flat terrain and (ii) the characteristics of the layer, in which atmospheric variables, such as humidity, heat or aerosol content, show a diurnal variation, deviate from those of the classical CBL. These differences are most likely caused by the impact of the Earth's surface by means of convection and mesoscale thermally driven circulations. In Sect. 2.1, the ABL was defined as that part of the atmosphere which is directly affected by the presence of the Earth's surface, responds to surface forcing within a time scale of around an hour and exhibits a diurnal variation (Stull, 1988). Considering response time scales longer than one hour, a mountain ABL can be defined as the layer that develops due to the impact of orography in the course of a day. This definition is in agreement with the concept presented in previous studies (e.g. De Wekker, 2002; Serafin and Zardi, 2010a). According to these time scales, convection, thermally driven circulations, such as slope winds, valley winds and sea breezes including their return flows and compensating subsidence, topographic venting and advective venting are regarded as boundary-layer processes. It can be assumed that different processes are dominant in the diverse parts of the mountain ABL.

2.4. Impact of the Background Flow on the Mountain ABL over Complex Terrain

The background flow is driven by large-scale pressure gradients, which occur on a horizontal length scale of several hundreds or more kilometres. Especially a strong background flow can impact and modify the thermally driven circulations and the structure of the mountain ABL over complex terrain via turbulent transport and due to dynamically driven flows, which occur when the background flow interacts with orography. The latter include gravity waves and downslope windstorms.

2.4.1. Turbulent Transport

As long as the background flow is weak and a strong inversion tops the valley atmosphere, the vertical exchange between the background flow and the valley atmosphere is limited (Whiteman, 2000). When the background flow is strong, wind shear erodes the inversion at the top of the valley atmosphere and the valley atmosphere gets coupled with the background flow. Momentum, heat and humidity from above can suddenly break into the valley atmosphere. This may result in rapid changes in temperature, humidity and wind due to turbulent downward transport and in an interruption of the thermally driven circulations. This transport mechanism is most likely to occur in an unstably or neutrally stratified atmosphere in the valley (e.g. Whiteman and Doran, 1993).

To determine if buoyancy or mechanical processes dominate the production of turbulence, the gradient Richardson number is often used (e.g. Stull, 1988)

$$Ri(z) = \frac{g}{\overline{\Theta}(z)} \frac{\frac{\partial \overline{\Theta}(z)}{\partial z}}{\left(\frac{\partial \overline{u}(z)}{\partial z}\right)^2 + \left(\frac{\partial \overline{v}(z)}{\partial z}\right)^2}, \tag{2.6}$$

which is the ratio between buoyant and mechanical production of turbulence. In an unstably stratified layer, turbulence may be generated by both production processes and Ri is negative. In a stably stratified layer, the restoring force of gravity acts against turbulent vertical motions. While turbulence is suppressed by negative buoyancy, wind shear produces turbulence mechanically. When Ri is lower than the critical Richardson number, $Ri_c = 0.25$, mechanical production presumably prevails and a stably stratified laminar flow becomes turbulent. For example, Corsmeier et al. (1997) analysed downward transport processes of ozone in a nocturnal boundary layer, which are initiated by mechanically produced turbulence, due to strong wind shear caused by a low-level jet.

An inversion separating two layers, could be eroded by mechanically produced turbulence when $Ri < Ri_c$ in the inversion, i.e. when the wind shear is strong or the inversion is weak enough. This erosion can be explained using Kelvin-Helmholtz waves that form in the inversion (e.g. Stull, 1988). When Ri in an initially laminar flow falls below Ri_c, gentle waves begin to form. The wave amplitudes keep growing, reaching a point where they begin to break. Because air with lower density has been transported below air with higher density within each wave, buoyant production of turbulence occurs. Both production processes cause a mixing of the different layers at the inversion, transferring momentum, heat and humidity between the layers. The formerly sharp transition zone becomes broader with weaker wind shear and static stability. Consequently, the layer with the inversion and wind shear is then shifted to a lower level. If $Ri < Ri_c$ persists, vertical mixing produced by mechanical production of turbulence may result in a neutrally stratified layer where conserved atmospheric quantities, like potential temperature or specific humidity, are about constant with height and which is established from top to bottom.

In the case of opposing flows in the two layers, the flow in the lower layer may decelerate or even reverse its sign (Adler et al., 2011a). As the humidity in the upper layer is typically lower and the potential temperature and the horizontal wind speed are higher, the turbulent downward transport results in a temperature and wind speed increase and a humidity decrease in the lower layer.

2.4.2. Dynamically Driven Flows

To understand the complex behaviour of dynamically driven flows in mountainous terrain, two theories are often applied: the linear gravity wave theory and the hydraulic flow theory.

Linear Gravity Wave Theory

When an air mass approaches a mountain it is forced to lift. To describe the ability of the air mass to pass over the mountain, the non-dimensional mountain height is often applied (e.g. Markowski and Richardson, 2010):

$$\hat{H} = \frac{NH}{U} \tag{2.7}$$

with the barrier height, H, the horizontal upstream wind speed, U and the Brunt-Väisäla frequency

$$N = \sqrt{\frac{g}{\overline{\Theta}(z)} \frac{\partial \overline{\Theta}(z)}{\partial z}}. \tag{2.8}$$

As $\hat{H}$ is typically calculated for a vertical layer, U and N are mean values averaged over the layer. Strong stable stratification or low wind speed on the upstream side lead to $\hat{H} > 1$, which implies that at least some depth of the flow is blocked by the mountain. When air flows above a mountain in a stably stratified atmosphere, internal gravity waves occur as a result from buoyancy forces, which affect air parcels when they are vertically displaced from their equilibrium level. Because internal gravity waves are initiated by orography, they must carry energy upwards away from the source and they are propagating in the horizontal and vertical.

Here, only the fundamental basics of the linear internal gravity theory are introduced. A detailed description is given by e.g. Markowski and Richardson (2010) and Nappo (2013). It is assumed that the atmospheric variables consist of a mean part, which is either constant or varies only with height, and a fluctuating part, which is smaller than the mean part and varies in space and time. For example, the vertical wind speed is written as

$$w = \overline{w} + w'. \tag{2.9}$$

These variables are then substituted into the equations of motions ignoring non-linear terms, i.e. any terms containing multiples of fluctuating parts. With this linearization, some aspects of the gravity waves are removed, like steepening and breaking of waves. To simplify the analysis, a two-dimensional Boussinesq flow in the x-z plain is assumed. Furthermore, the mean vertical wind speed, $\overline{w}$, is set to zero so that $w = w'$. This set of equations is then reshaped into one equation for w and an ansatz for one mode of w is assumed to be

$$w = \tilde{w}(z)\ e^{i(kx-\omega t)}, \tag{2.10}$$

where $\tilde{w}$ is a complex amplitude, which depends on height; k is the wavenumber in x-direction; and $\omega(z)$ is the angular frequency in a fixed reference frame. Substituting this into the equation for w and neglecting the effect of decreasing density with height on the wave amplitude leads to the Taylor-Goldstein equation

$$\frac{d^2\tilde{w}}{dz^2} + \left(\frac{k^2N^2}{\Omega^2} + \frac{k}{\Omega}\frac{d^2\overline{u}}{dz^2} - k^2\right)\tilde{w} = 0 \tag{2.11}$$

with the intrinsic frequency

$$\Omega = \omega - \overline{u}k, \tag{2.12}$$

which the wave would have when observed in a reference frame moving with the mean wind, $\overline{u}$. Assuming a steady state, i.e. $\omega = 0$, Eq. 2.11 results in

$$\frac{d^2\tilde{w}}{dz^2} + \left(\frac{N^2}{\overline{u}^2} - \frac{1}{\overline{u}}\frac{d^2\overline{u}}{dz^2} - k^2\right)\tilde{w} = \frac{d^2\tilde{w}}{dz^2} + \left(l^2 - k^2\right)\tilde{w} = 0 \tag{2.13}$$

with the Scorer parameter

$$l = \sqrt{\frac{N^2}{\overline{u}^2} - \frac{1}{\overline{u}}\frac{d^2\overline{u}}{dz^2}}. \tag{2.14}$$

Equation 2.10 is then written as

$$w = Ae^{i(kx+mz)} + Be^{i(kx-mz)} \tag{2.15}$$

with the vertical wavenumber

$$m = \sqrt{l^2 - k^2} \tag{2.16}$$

and complex wave amplitudes A and B. The height dependency of w is determined by whether m is real or imaginary, thus it depends on whether $l^2 > k^2$ or not. If m is real, horizontally and vertically propagating two-dimensional waves are possible. If m is imaginary, w exhibits exponential decay with height and is called an evanescent solution. Thus, the wave characteristics are determined by the height dependency of l. When $l^2 > k^2$ in a lower layer and $l^2 < k^2$ in an upper layer, i.e. l decreases with height, wave energy can become trapped within the lower layer and trapped lee waves develop (e.g. Jackson et al., 2013).

In a simplified environment, $\overline{u} = U$ as the mean upstream wind speed and N are assumed to be constant with height. Equation 2.16 then leads to

$$m^2 = \frac{N^2}{U^2} - k^2 = \frac{k^2 N^2}{\Omega^2} - k^2. \tag{2.17}$$

For m to be real and thus to allow for wave motions, $\Omega^2 < N^2$ is necessary. Waves with $\Omega^2 > N^2$ are evanescent. Thus, the oscillation frequency of waves is limited by N. This implies that the smallest possible horizontal wavelength for steady waves is given by (Stull, 2000)

$$\lambda_{\text{min}} = \frac{2\pi \cdot U}{N}. \tag{2.18}$$

Besides the vertical structure of the atmosphere, internal gravity waves characteristics may also depend on the underlying orography. Gravity waves with a horizontal wavelength of about twice the mountain-ridge width may be amplified resulting in wave amplitudes of roughly half the mountain-ridge height (Stull, 2000). When gravity waves are initiated by successive ridges, they can either be amplified or damped depending on the relationship between the ridge separation distance and the horizontal wavelength (Gerbier and Berenger, 1961). If the distance between the ridges

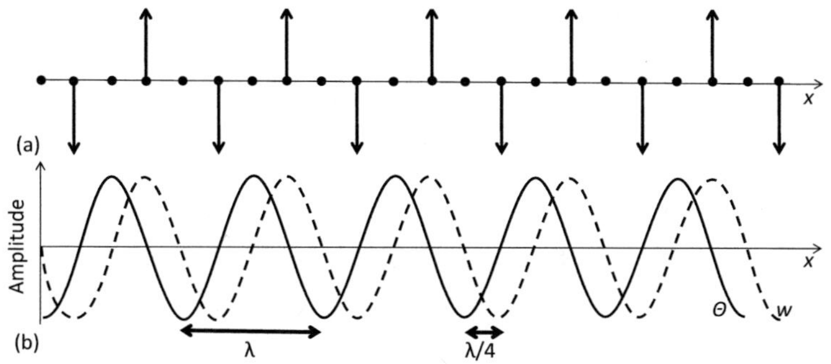

Figure 2.4.: Schematic diagram of the instantaneous vertical motion of air parcels in a horizontally propagating internal gravity wave with $U = 0$ m s^{-1} (a) and corresponding fluctuations of potential temperature, Θ, and vertical wind speed, w, with a horizontal wavelength, λ, in x-direction (b).

is a multiple of the horizontal wavelength, the wave amplitude may be intensified. The wavelength which is most favourable for large-amplitude waves is given by the so-called terrain wavelength. Thus, when the wavelength of the gravity wave matches the terrain wavelength resonance may result in large-amplitude waves.

To identify gravity waves in aircraft data, it is useful to determine the phase shift in the fluctuations of potential temperature and vertical wind speed. For gravity waves this phase shift is 90° (Durran, 1990). This is a distinct characteristic of gravity waves and can be used to distinguish between gravity waves and convection, for which fluctuations of potential temperature and vertical wind speed are in phase. Thus, an aircraft, flying on a horizontal flight track through a gravity wave, would measure variations of potential temperature and vertical wind speed, which are shifted by a quarter of the horizontal wavelength, i.e. they are 90° out of phase (Fig. 2.4). When the fluctuations of potential temperature are largest, i.e. at the air parcels maximum positive or negative displacement, the vertical wind speed is zero. Between these locations the air parcels are either gaining or loosing buoyancy, reaching their maximum vertical wind speed when variations of potential temperature are smallest.

Hydraulic Flow Theory

Assuming that stratified air flowing over a mountain is analogous to a water flow in a channel, the hydraulic flow theory, first presented by Long (1954), can be used for a physical explanation of dynamically driven flows. In the following, the hydraulic flow theory is briefly described for a fluid layer with a free upper surface, which flows over an obstacle of height, H, e.g. water in a river that flows over a rock. Based on a one-dimensional version of the shallow water equations and assuming a steady flow without friction, external pressure gradients and Coriolis force, the nonlinear momentum and continuity equations can be written as (e.g. Markowski and Richardson, 2010)

$$u\frac{\partial u}{\partial x} + g\frac{\partial D}{\partial x} = -g\frac{\partial H}{\partial x} \tag{2.19}$$

$$\frac{\partial}{\partial x}(uD) = 0 \tag{2.20}$$

with the fluid speed, u, in x-direction being height-constant throughout the fluid layer depth, D. Combining Eqs. 2.19 and 2.20, results in

$$\left(1 - \frac{u^2}{gD}\right)\frac{\partial D}{\partial x} = -\frac{\partial H}{\partial x}. \tag{2.21}$$

With the shallow water phase speed

$$c \equiv \sqrt{gD}, \tag{2.22}$$

the Froude number for hydraulic flow theory in case of a free upper surface is defined as

$$Fr = \frac{u}{c} = \frac{u}{\sqrt{gD}}. \tag{2.23}$$

In a supercritical flow ($Fr > 1$), u is larger than c and no gravity waves at the free upper surface can propagate upstream relative to the mean flow. The fluid deepens and decelerates as it goes up the upstream slope, reaches its maximum thickness at the peak and thins and accelerates again going down the downstream slope. When $Fr < 1$, so-called subcritical flow, gravity waves at the free surface can propagate up- and downstream. When going up the upstream slope, the fluid thins and accelerates, reaching its minimum thickness and maximum speed at the peak. Going down the downstream slope, it thickens and decelerates.

To gain a net acceleration of the fluid at the bottom of the downstream slope, a transition from subcritical flow on the upstream side to supercritical flow on the downstream side is necessary. Thus, a Froude number of $Fr = 1$ around the ridge has to occur. This means that the fluid speed at the mountain ridge, u_{ridge}, has to equal c. Further downstream, the fluid eventually decelerates and thickens in a hydraulic jump, associated with the generation of strong turbulence. Assuming that and in accordance with the conservation of energy, applying the Bernoulli equation

$$\frac{1}{2}u^2 + gD = gD_{\text{upstream}} \tag{2.24}$$

with D_{upstream} as the far upstream fluid depth, yields a decrease of the fluid depth by $\frac{2}{3}$ at the ridge compared to the far upstream conditions (e.g. Jackson et al., 2013). The conservation of mass then implies an acceleration of the flow at the ridge to $u_{\text{ridge}} = \frac{3}{2}U$.

As free surfaces do not exist in the continuously stratified atmosphere, some difficulties arise when applying the hydraulic flow theory to the atmosphere (e.g. Durran, 2003). The calculation of precise quantitative values for Fr turns out to be difficult in the real, stratified atmosphere with indistinct inversions and when wind speed and stratification are not height constant within the different layers. Nevertheless, the atmosphere is often separated into two layers with constant potential temperature and wind speed and an inversion in between. For a two-layer atmosphere, the Froude number from Eq. 2.23 is modified to

$$Fr = \frac{u}{\sqrt{gD\frac{\Delta\Theta}{\Theta_{\text{m}}}}}, \tag{2.25}$$

with $\Delta\Theta$ being the potential temperature difference across the inversion and Θ_{m} being the temperature of the lower layer. The denominator describes the phase speed of internal gravity waves that form within the inversion.

Downslope Windstorms

Downslope windstorms modify the atmosphere downstream of the mountain ridge by transporting air masses from higher layers downwards. Foehn, which is a type of a downslope windstorm, is characterized by warm and dry air and high wind speed on the downstream side of the mountain ridge (e.g. Brinkmann, 1971). To explain the existence of a downslope windstorm, various approaches exist, based on linear gravity wave theory or hydraulic flow theory (e.g. Jackson et al., 2013; Richner and Hächler, 2013). In terms of linear gravity wave theory, these approaches include the amplification of vertically propagating gravity waves, by waves reflected off pre-existing critical layers (Klemp and Lilly, 1975) or reflected off a self-induced critical layer created by wave breaking (Peltier and Clark, 1979). In linear theory, vertically propagating gravity waves can get reflected off layers where the Scorer parameter changes rapidly with height due to changes in temperature or wind speed (e.g. Markowski and Richardson, 2010). On the basis of hydraulic flow theory, a downslope windstorm occurs when the Froude number in the flowing layer equals one at the mountain ridge.

Interactions between Dynamically Driven Flows and the Downstream ABL

In most previous studies the effect of the ABL on dynamically driven flows was ignored (e.g. Smith, 1980). More recent modelling studies demonstrated an impact of the ABL on dynamically driven flows via a damping effect which reduces the wave amplitudes, the strength of downslope windstorms and the tendency for wave breaking (e.g. Richard et al., 1989; Smith, 2007; Vosper and Brown, 2007; Jiang et al., 2008). Dynamically driven flows were subject of several field campaigns in the past. For example, subgoals of the MAP campaign (Bougeault et al., 2001) focused on Alpine foehn and the T-REX campaign (Grubišic et al., 2008) in the Sierra Nevada in California aimed to investigate downslope windstorms and rotors.

A common finding was that the ABL interacts with dynamically driven flows and thus must not be neglected (e.g. Gohm et al., 2004; Zängl et al., 2004). Observations in the Alps revealed the attenuation of a stationary gravity wave due to a stagnant layer with weak winds and stable stratification at lower levels (Smith et al., 2002). Smith et al. (2006) proposed a two-dimensional theoretical boundary-layer model to analyse the impact of the

ABL on the absorption of gravity waves and found that the ABL attenuates the wave and shortens its wavelength. Based on numerical simulations and theoretical formulations, Jiang et al. (2006) investigated the impact of the ABL on trapped waves. They found that the absorption of wave energy is higher for a stable boundary layer than for a CBL and suggest that an absorbing ABL serves as an energy sink for trapped waves. The interaction between nocturnal thermally driven downslope winds and gravity waves is highly complex and non-linear, as was described by Poulos et al. (2000, 2007) using observations and simulations.

During foehn conditions, a downstream cold air pool inhibits the penetration of the foehn air mass to the ground (Gubser and Richner, 2001; Vogt and Jaubert, 2004). Heating from below due to the sensible surface heat flux as well as turbulent entrainment induced by Kelvin-Helmholtz waves at the top of the cold pool eventually erode the nocturnal inversion (Nater et al., 1979) and allow the foehn air mass with a low aerosol content to progressively reach the ground (Frioud et al., 2004). Flow blocking on the windward site reduces the wave amplitude due to a reduction in the effective barrier height (Jiang et al., 2005). Jiang and Doyle (2008) studied the diurnal variations of cross-valley flows based on observations and numerical simulations and found that dynamically driven downslope winds strongly depend on the thermal structure of the downstream valley atmosphere. Mayr and Armi (2010) investigated how diurnal heating warms the downstream valley atmosphere and allows the flow crossing the mountain ridge to descend into the potentially warmer valley atmosphere.

The extension of a downslope windstorm or the amplitude of lee waves may depend on the downstream stratification, as indicated by the above described previous studies. A reduction of downstream stability may therefore allow the downslope windstorm to extent further towards the ground downstream of the mountain ridge and over a longer horizontal distance. The simple large-amplitude lee wave theory assumes that the downstream stratification behind the mountain ridge is the same as upstream. When the downstream stratification is less stable than on the upstream side, the wave amplitude may be enhanced on the downstream side, compared to the case with equal stratification on both sides of the mountain ridge. This may affect the wavelength and amplitude of the lee wave. As the intruding air mass is typically potentially warmer and drier and has a higher momentum than the air mass in lower layers, a lower downstream stability may lead to warmer and drier air and higher wind speeds close to the ground over long distances downstream of the mountain ridge.

2.5. Atmospheric Water Vapour and its Detection

The atmospheric water-vapour content can be described by the specific humidity, q, which is defined as the ratio of water-vapour density, ρ_v, to the density of moist air, ρ_m, and is approximated as (e.g. Kraus, 2004)

$$q = \frac{\rho_v}{\rho_m} \approx 0.622\frac{e}{p}, \tag{2.26}$$

where e is the water-vapour pressure and p is the air pressure. Instead of ρ_m the density of dry air, ρ_d, may be used together with the virtual temperature

$$T_v = T(1 + 0.61q) \tag{2.27}$$

that accounts for the presence of water vapour in the air. Other humidity variables are the water-vapour mixing ratio

$$r_v = \frac{\rho_v}{\rho_d} \tag{2.28}$$

and the dewpoint temperature, T_d, which is the temperature at which saturation is achieved if air is cooled while holding p and r_v constant (e.g. Markowski and Richardson, 2010):

$$T_d = \frac{243.5}{\left(\frac{17.67}{\ln\left(\frac{e}{6.112}\right)}\right) - 1} \tag{2.29}$$

The relative humidity describes the saturation of an air mass and depends on air temperature and pressure. It is given by

$$rh = \frac{e}{e_s}100\ \%, \tag{2.30}$$

where e_s is the saturation water-vapour pressure, which can be approximated as (e.g. Bolton, 1980):

$$e_s = 6.112\ \mathrm{e}^{\frac{16.67T}{T+243.5}}. \tag{2.31}$$

The prognostic equation for the conserved mean specific humidity, $\overline{q}$, in a turbulent flow is (e.g. Stull, 1988)

$$\underbrace{\frac{\partial \overline{q}}{\partial t}}_{\text{I}} = \underbrace{-\overline{u}\frac{\partial \overline{q}}{\partial x} - \overline{v}\frac{\partial \overline{q}}{\partial y} - \overline{w}\frac{\partial \overline{q}}{\partial z}}_{\text{II}} \underbrace{- \frac{\partial \overline{u'q'}}{\partial x} - \frac{\partial \overline{v'q'}}{\partial y} - \frac{\partial \overline{w'q'}}{\partial z}}_{\text{III}} \underbrace{+ \nu_q \nabla^2 \overline{q}}_{\text{IV}} \underbrace{+ \frac{\overline{M}}{\rho_d}}_{\text{V}}, \tag{2.32}$$

with the mean wind components $\overline{u}$, $\overline{v}$ and $\overline{w}$ in x-, y- and z-direction; the fluctuating wind components u', v' and w'; the kinematic molecular diffusivity for water vapour, ν_q; and the mass of water vapour per unit volume and time, which is created by phase changes from liquid or solid to vapour and vice versa, $\overline{M}$. Term I represents the storage of $\overline{q}$ and results from the contributions of terms II-IV. Term II describes the advection of $\overline{q}$ by the mean wind, term III illustrates the divergence of the turbulent kinematic humidity flux, term IV represents the mean molecular diffusion of water vapour and term V depicts the contribution due to phase changes. As a flux is the transfer of a quantity per unit area and time, term III describes the turbulent transport of humidity and contributes to an increase or decrease of the mean humidity depending on whether there is a flux convergence or divergence. The latent heat flux, E, is associated with the vertical part of term III

$$E = \overline{\rho}_d L \overline{w'q'}, \tag{2.33}$$

where $\overline{\rho}_d$ is the mean density of dry air and L is the specific latent heat of vaporization. At the Earth's surface, E describes the transport of humidity from and to the surface. When it is positive, it represents the part of the net radiation that is used for evapotranspiration from soil and vegetation. Besides the turbulent transport, advection with the mean wind significantly impacts the humidity distribution in the atmosphere. This leads to a very high spatial variability of humidity, especially in the ABL (e.g. Weckwerth et al., 1996; Khodayar, 2009).

To measure atmospheric humidity different methods exist. Sensors measuring the humidity in-situ provide time series when they are equipped at surface stations or profiles when they are launched with radiosondes. The advantage of these methods are the high temporal and vertical resolution, respectively. However, a major shortcoming is that both measurements are point measurements or instantaneous and do not represent a large area. In addition, radiosondes are not launched continuously and thus the temporal resolution of the humidity profiles can be very low. Khodayar (2009) found that the high spatial variability of humidity in the ABL cannot be resolved with radiosonde data only.

The IWV describes the total water-vapour density in an air column with an unit area reaching from the Earth's surface, h_0, to the top of the troposphere, h_t:

$$\mathrm{IWV} = \int_{h_0}^{h_t} \rho_v(z)\mathrm{d}z. \tag{2.34}$$

Because the water-vapour density has an average scale height of about 2000 m (e.g. Bock et al., 2005), variability of IWV is closely related to the humidity in the lower troposphere. Integrating the prognostic equation for the mean humidity (Eq. 2.32) over a vertical column, it is evident that the total atmospheric water-vapour content in an air column depends on evapotranspiration from the Earth's surface, on horizontal advection and on phase changes, neglecting the horizontal divergence of the turbulent kinematic humidity fluxes and molecular diffusion and assuming a vanishing vertical wind speed at the bottom and top of the vertical air column. Without any advection and phase changes, evapotranspiration or dew formation at the Earth's surface lead to an increase or decrease of water-vapour content in the vertical air column and an increase or decrease of IWV. Horizontal advection of moist or dry air in any layer of the air column contributes to an increase or decrease of IWV. Phase changes from solid or liquid into vapour or vice versa increase or decrease the IWV. This occurs when clouds dissolve or form in the air column.

In the past, several methods to determine IWV have been applied. Remote sensing methods include e.g. ground-based microwave radiometry or a technique using the ground-based Global Positioning System (GPS) network.

Ground-based Microwave Radiometry

The first method described here is the ground-based microwave radiometry, which is based on passive remote sensing. The basic concepts are given by e.g. Janssen (1993) and Crewell et al. (2011). The microwave region covers electromagnetic radiation with wavelengths between 1 mm and 10 cm, i.e. they include frequencies from 300 to 3 GHz. A microwave radiometer measures the thermal emission along the wings of pressure-broadened rotational lines of oxygen and water vapour (Fig. 2.5). As oxygen is homogeneously mixed throughout the troposphere, measurements along the low frequency wing of the oxygen absorption complex at 60 GHz provide information on vertical temperature profiles (Löhnert et al., 2009). Close to the centre of the complex the atmosphere is opaque and most of the information originates from close to the surface. Towards lower frequencies the atmosphere becomes less opaque which allows information from higher atmospheric layers to be detected. Humidity information are obtained from the optically thin water-vapour line at 22.235 GHz. Microwave radiation can penetrate through clouds, which allows measurements in the cloudy atmosphere. Only precipitation contaminates the detected signals (Rose et al., 2005).

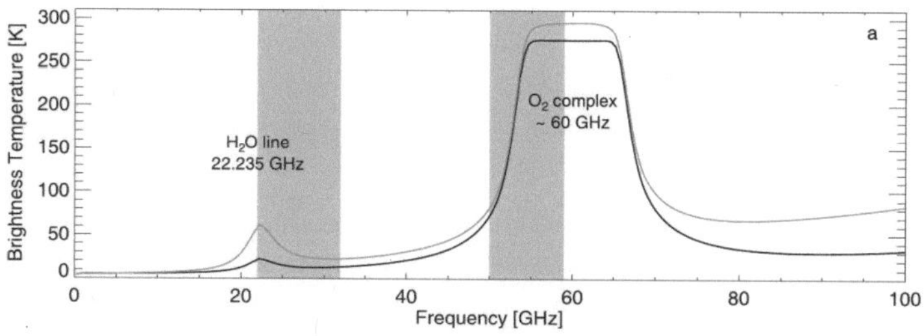

Figure 2.5.: Microwave spectrum during a summer day (grey line, IWV≈ 30 kg m^{-2}) and during a winter day (black line, IWV≈ 8 kg m^{-2}) at Payerne (Löhnert et al., 2009). Grey shadings mark the parts of the spectrum used for humidity and temperature profiling.

The measurement principle is based on the radiative transfer in the troposphere. The differential form of the radiative transfer equation can be written as (Janssen, 1993)

$$\frac{dI_f}{ds} = -I_f\alpha + S_f, \tag{2.35}$$

where I_f is the radiant power per unit area and per unit solid angle within a specified interval of the frequency, f; s is the path of propagation; α is the absorption coefficient; and S_f is a source term. In the absence of scattering and according to Kirchoff's law the source term equals

$$S = \alpha B_f(T) \tag{2.36}$$

with the Planck function

$$B_f(T) = \frac{2hf^3}{c_0^2}\frac{1}{e^{\frac{hf}{k_B T}} - 1}, \tag{2.37}$$

where h is the Planck constant; c_0 the speed of light in vacuum; k_B the Boltzmann constant; and T the physical temperature. $B_f(T)$ describes the quantity of radiation that passes through a unit surface within a given solid angle and frequency interval. This feature allows the radiation transfer along a pencil beam to be measured with a radiometer. The downward spectral radiance, $I_f(h_0)$, received by a ground-based radiometer at height h_0 can be obtained by solving the radiative transfer equation (Eq. 2.35) with use of Eq. 2.36 as

$$I_f(h_0) = I_f(\infty)\mathrm{e}^{-\tau(\infty)} + \int_{h_0}^{\infty} \alpha(s)B_f(T(s))\mathrm{e}^{-\tau(s)}\mathrm{d}s, \tag{2.38}$$

where the optical depth of the atmosphere is defined as

$$\tau(s) = \int_{h_0}^{s} \alpha(s')\mathrm{d}s'. \tag{2.39}$$

In Eq. 2.38, $I_f(\infty)$ represents the spectral radiance of the cosmic background and $\mathrm{e}^{-\tau(\infty)}$ describes the overall transmission of the atmosphere. In general, the first term of Eq. 2.38 is much smaller than the second one. In the microwave region, the Rayleigh-Jeans-Approximation applies, as $hf \ll k_B T$, and a linear relation between $B_f(T)$ and T results from the Planck function (Eq. 2.37). With the definition of the brightness temperature

$$T_B(f) \equiv \frac{c_0^2}{2f^2 k_B} I_f \tag{2.40}$$

$T_B(f)$ observed by a ground-based radiometer can be obtained from Eq. 2.38:

$$T_B(f) = T_{B\infty} \mathrm{e}^{-\tau(\infty)} + \int_{h_0}^{\infty} \alpha(s) T(s) \mathrm{e}^{-\tau(s)} \mathrm{d}s, \tag{2.41}$$

where $T_{B\infty}$ is the brightness temperature of the cosmic background. To describe the height where the signal originated, a weighting function is defined, which generally varies with height and depends on α (Crewell et al., 2011). Since the pressure broadening of the spectral lines depends on the pressure (2 to 3 MHz hPa^{-1}), frequencies ranging over several GHz can be used to obtain tropospheric profiles. For example, for frequencies close to the centre of the water-vapour line the weighting function increases with decreasing pressure, i.e. layers at higher elevations contribute more to the measured T_B. At the same time, the contribution from layers at lower elevations is larger at the wings due to higher weighting functions. The weighting function at about 24 GHz is approximately height-constant, which makes it favourable for the determination of IWV.

To obtain information on atmospheric variables from T_B, an inversion of the radiative transfer equation is necessary. Statistical methods are based on regression models, e.g. linear or neural networks, which link the observed T_B to the atmospheric variables.

Water-Vapour Detection with GPS

Secondly, IWV can be derived from GPS signals (e.g. Bevis et al., 1992; Wickert and Gendt, 2006). The great advantage of using IWV from GPS measurements is the rather dense network of GPS receivers all over the world. GPS satellites at an altitude of 20200 km transmit microwave signals that are picked up by ground-based GPS receivers. The GPS signal is phase-delayed through refractive effects in the ionosphere and troposphere compared to its propagation in the vacuum. The dispersive ionospheric delay is in general removed by a linear combination of dual-frequency observations.

The tropospheric delay (Δtrop) depends on pressure, temperature and humidity. Because even small amounts of water vapour significantly impact Δtrop (Rocken et al., 1993), the GPS technique is suitable to detect tropospheric water vapour. To obtain the Zenith Total Delay (ZTD) from Δtrop, mapping functions are used. It is assumed that ZTD has two components: the Zenith Wet Delay (ZWD) caused by the atmospheric water vapour and the Zenith dry or Hydrostatic Delay (ZHD) arising from the dry air:

$$\mathrm{ZTD} = \mathrm{ZHD} + \mathrm{ZWD}. \tag{2.42}$$

The impact of liquid water on ZWD cannot be extracted from GPS measurements and is ignored. The arising error is typically less than 5 % (Solheim et al., 1999). By the use of mapping functions $\mathrm{m_{hyd}}$ for ZHD and $\mathrm{m_{wet}}$ for ZWD, the tropospheric delay results from

$$\Delta\mathrm{trop} = \mathrm{ZHD}\ \mathrm{m_{hyd}} + \mathrm{ZWD}\ \mathrm{m_{wet}}. \tag{2.43}$$

Depending on the surface pressure, p_0, at the GPS receiver and the latitude, φ, of the station, ZHD can be calculated (Bock et al., 2007):

$$\mathrm{ZHD} = 2.279(\mathrm{mm\ hPa^{-1}})\frac{p_0}{f(\varphi, h_0)}, \tag{2.44}$$

where $f(\varphi, h_0)$ describes the correction of the gravity field (Saastamoinen, 1972). ZWD can then be derived from Eq. 2.43 and ZTD from Eq. 2.42. Simultaneous measurements from one GPS receiver to several GPS satellites allow to calculate ZTD with an accuracy of a couple of millimetres. The zenith IWV is roughly proportional to ZWD with the conversion factor $\kappa(T_\mathrm{m})$:

$$\mathrm{IWV} = \kappa(T_\mathrm{m})\ \mathrm{ZWD}, \tag{2.45}$$

where T_m is the water-vapour weighted mean temperature in the atmospheric column above the GPS antenna. As a rule of thumb, a conversion factor of $\kappa(T_\mathrm{m}) \approx 0.15$ applies, however it can vary up to 20 % as T_m depends on local conditions (Bevis et al., 1994). T_m can either be modelled as a linear function of surface temperature or obtained from numerical weather models. As these data usually have to be interpolated or extrapolated additional error sources arise and the preferred use of surface observations was suggested (Hagemann et al., 2003; Bock et al., 2005).

2.6. Convection-Related Parameters

Besides the existence of a trigger mechanism and a sufficient amount of humidity in the lower and middle troposphere, some kind of atmospheric instability is required for the evolution of deep convection. An unstable stratification of the environmental air enhances the vertical motion of a perturbed air parcel, while a stable stratification suppresses the vertical motion. To evaluate the stability of an unsaturated atmosphere with respect to the development of clouds, the parcel theory is often applied, which describes the adiabatic ascent of an air parcel from a certain layer, usually close to the surface. This theory is able to assess the existence of conditional instability, i.e. the atmosphere is locally stable to the displacement of unsaturated air parcels, but unstable to the displacement of saturated air parcels (e.g. Emanuel, 1994). Although this theory neglects pressure perturbations and assumes that condensate is removed from the air parcel as soon as it forms, it is useful to assess an upper bound of the energy available for moist convection. Another shortcoming of the parcel theory is the neglect of the exchange of momentum, humidity and heat between the parcel and the environmental air. Entrainment of environmental air into a rising parcel reduces its buoyancy and slows down the ascent of the parcel (e.g. Markowski and Richardson, 2010). This process is not considered in the CAPE values, predicted by parcel theory, but may have a significant impact on the evolution of moist convection.

To analyse the characteristics of the tropospheric vertical profile, to visualize the ascent of an air parcel and to derive various unmeasured parameters, thermodynamic diagrams are widely used. A common thermodynamic diagram is the Stüve diagram (Fig. 2.6). On the ordinate $p^{\frac{R_d}{c_p}}$ (R_d: gas constant for dry air, c_p: specific heat at constant pressure for dry air) is shown, which allows to display a wide pressure range. An unsaturated air parcel that is vertically displaced maintains its potential temperature and follows a dry adiabat, which are straight lines in a Stüve diagram. The dry adiabats intersect at $T = -273.15$ °C and $p = 0$ hPa. The temperature of a vertically displaced saturated air parcel decreases with the pseudoadiabatic lapse rate (e.g. Markowski and Richardson, 2010), assuming that all condensate is immediately removed from the parcel as it forms. This air parcel follows a pseudoadiabat,

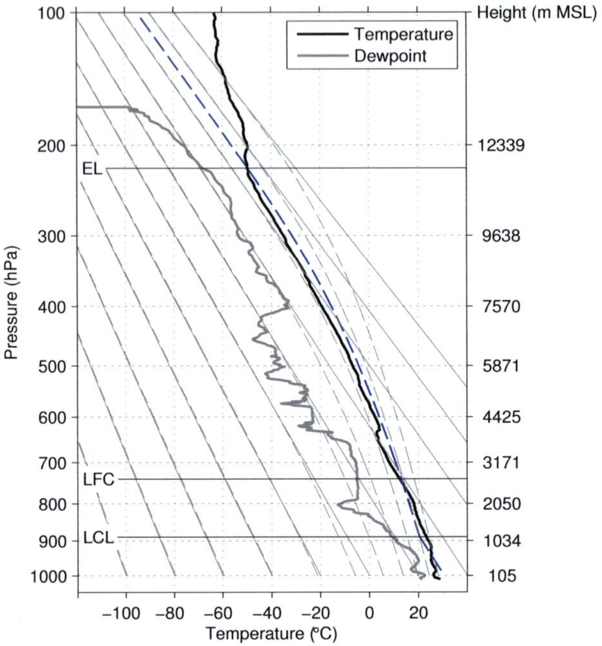

Figure 2.6.: Stüve diagram with temperature and dewpoint obtained from a radiosounding at Ajaccio at 1300 LT on 25 August 2012. The dashed blue line indicates the virtual temperature of an air parcel with the mean layer temperature and humidity values of the lowest 50 hPa, which is lifted from 25 hPa above the surface to its Lifting Condensation Level (LCL), Level of Free Convection (LFC) and Equilibrium Level (EL). The thin solid grey lines indicate dry adiabats and dashed grey lines pseudoadiabats.

which approaches a dry adiabat for low temperatures. The air parcel is lifted dry adiabatically by either buoyant or mechanical forces to its Lifting Condensation Level (LCL). Because the potential temperature and mixing ratio are conserved during a dry adiabatic ascent, the LCL can be determined as the intersection of the dry adiabat and the constant saturation mixing ratio line of the parcel to be lifted. Above the LCL, the condensation of water vapour within the parcel releases latent heat, which warms the parcel and results in the lower pseudoadiabatic lapse rate. When the lifting is strong enough, the air parcel can reach its Level of Free Convection (LFC) and its density becomes lower than the density of the environmental air. The air parcel is then positively buoyant and rises till its Equilibrium Level (EL), where its density becomes higher than the one of the environmental air. The area bounded by the environmental temperature profile, T_v, and the pseudoadiabat, T_{v_p}, followed by a saturated air parcel rising from its LFC to EL represents the total amount of potential energy available for convection (e.g. Markowski and Richardson, 2010)

$$\mathrm{CAPE} = \int_{\mathrm{LFC}}^{\mathrm{EL}} g \, \frac{T_{v_p} - T_v}{T_v} \mathrm{d}z. \tag{2.46}$$

The area between the environmental temperature profile and the path taken by a negatively buoyant air parcel below its LFC illustrates the energy that is necessary to initiate convection

$$\mathrm{CIN} = - \int_{h_0}^{\mathrm{LFC}} g \, \frac{T_{v_p} - T_v}{T_v} \mathrm{d}z. \tag{2.47}$$

In this thesis, an air parcel with the mean values of temperature and humidity of the lowest 50 hPa is lifted, i.e. h_0 in Eq. 2.47 is 25 hPa above the surface. The pre-convective values of CAPE and CIN in an environment leading to the evolution of deep convection may vary a lot depending on the location. For example, Kunz (2007) found moderate convection over southern Germany for CAPE values larger than approximately 500 J kg^{-1}. However, the various parameters and thresholds need to be verified and tuned for each area using the local climatology (e.g. Dalla Fontana, 2008).

An often observed feature in a pre-convective environment is a layer of warm, dry air, a so-called lid, which traps moist air in the lower troposphere beneath potentially colder air in the middle and upper troposphere (Carlson and Ludlam, 1968; Browning et al., 2007). The criteria for a lid include that it occurs below 500 hPa, the relative humidity decreases by more than 1 % hPa^{-1} near the lid and the temperature increases with height or is height-constant (Graziano and Carlson, 1987). In general, lids form when warm, dry air is advected over a moist lower layer that is related to topographic features (Graziano and Carlson, 1987). While lids may suppress the release of conditional instability, they may also cause an increase of conditional instability with time and thus may enhance the likelihood for severe deep convection. For convection initiation the lid must be penetrated, which can either be caused by low-level warming and moistening or by mesoscale variability in the lid strength itself (Browning et al., 2007). Processes in the mountain ABL (Sect. 2.3) may favour the lid penetration over complex terrain.

To account for static stability and low-level humidity, the Total Totals (TT) index was developed (Miller, 1972). It combines two indices, the Vertical Totals index and the Cross Totals index. In this thesis, a modified TT index was used which considers the conditions at the surface and 850 hPa (Charba, 1979):

$$\mathrm{TT} = 0.5(T_{850} + T_{h_0}) + 0.5(T_{d,850} + T_{d,h_0}) - 2(T_{500}), \qquad (2.48)$$

where the indices 850, h_0 and 500 indicate air and dewpoint temperatures in 850 hPa, at the surface and in 500 hPa. The TT index has the advantage, in comparison with CAPE, that it is defined for each profile, provided that it reaches 500 hPa. Values higher than a threshold of 57 °C are assumed to be favourable for the evolution of deep convection (Ducrocq et al., 1998).

To release some kind of instability, a trigger mechanism is necessary. Over complex terrain this is provided by e.g. direct orographic lifting to the LFC or convergence zones generated by thermally driven circulations (Banta, 1990), leading to the initiation of moist (Meißner et al., 2007) or deep convection (Barthlott et al., 2006; Kalthoff et al., 2009). Over mountainous islands the superposition of various thermally driven circulations, like valley winds and sea breezes, enhance the convergence, which leads to a concentration of deep convection over islands in the afternoon and evening (Qian, 2008).

Several studies found that environmental water vapour played an important role in the evolution of deep convection (Brown and Zhang, 1997; Redelsperger et al., 2002; Kuang and Bretherton, 2006). Derbyshire et al. (2004) documented the sensitivity of the evolution of deep convection to environmental relative humidity in the middle troposphere and found that a lower humidity content in this layer is unfavourable for deep convection because of stronger evaporative cooling and negative buoyancy. Chaboureau et al. (2004) reported that a moistening of the atmosphere in the cloud layer is crucial for the subsequent evolution of deep convection. Semi-idealized simulations over the Corsican Island revealed a high dependency of the formation of deep convection on the humidity content in the middle troposphere (Metzger et al., 2014). These results suggest that the mid-level humidity should be included in the characterization of the pre-convective environment. For this purpose, the mean relative humidity (RH) was used in this thesis. It was averaged between 850 and 700 hPa (between about 1500 and 3000 m MSL):

$$\mathrm{RH} = \frac{1}{\Delta p} \int_{850\ \mathrm{hPa}}^{700\ \mathrm{hPa}} rh(p)\mathrm{d}p, \tag{2.49}$$

where $\Delta p = 150$ hPa and $rh(p)$ is the relative humidity profile.

In addition, water-vapour variability in the lower troposphere can have a significant impact on the evolution of moist convection (e.g. Weckwerth et al., 2004; Lascaux et al., 2004; Meißner et al., 2007). Crook (1996) reported that variations of specific humidity by 1 g kg^{-1} were decisive for the initiation of deep convection. Because CAPE and CIN strongly depend on the water-vapour content of the air parcel to be lifted (e.g. Adler et al., 2011b) and the TT index accounts directly for low-level humidity, a strong spatial variability of these parameters can be expected. Weckwerth (2000) stated that radiosoundings taken within the downward branches of horizontal convective roles in the ABL are not representative for the environmental stratification. Especially over complex terrain, large spatial differences in CAPE and CIN were reported (e.g. Kalthoff et al., 2009; Barthlott et al., 2010), which suggest strong spatial variability in the pre-convective environment.

3. The HyMeX Field Campaign and Data Base

3.1. The HyMeX Field Campaign

In order to improve the understanding of the initiation and development of heavy precipitation events in the Mediterranean region, the HyMeX project was initiated. Under this project, a Special Observation Period (SOP) was conducted in autumn 2012 to provide comprehensive data sets for process studies and model evaluation alike. Due to the processes involved in convective precipitation, observations on different scales were carried out. As one of the HyMeX's supersites, the KITcube was positioned on the Corsican Island in late summer and autumn.

In the framework of HyMeX, two main scientific goals were pursued with the deployment of this supersite: (i) the Corsican Island unites both features necessary to cause orographically induced convection or to modify it, i.e. a landmass of sufficient size (about 80 km by 180 km) and a high mountain ridge of up to 2710 m MSL, with about 20 mountains being higher than 2000 m MSL. The KITcube investigated the influence of the island on initiation and modification of isolated convection. (ii) For most of the synoptic conditions, under which severe precipitation events occur in the Ligurian Sea, the Corsican Island is on the upstream side (Lambert et al., 2011). Thus, the KITcube was well-positioned to provide upstream conditions for the intense precipitation events affecting continental south-eastern France and northern and central Italy.

3.2. Measurements on the Corsican Island during HyMeX

3.2.1. Terrain Characteristics and Station Locations

A main mountain ridge stretches across the island from the north-northwest to the south-southeast (Fig. 3.1). In the northern half of the island, the Tavignano Valley runs from Aléria on the east coast to Corte in the centre with a north-west to south-east oriented valley axis. At Corte, the valley axis bends towards north. Separated by the Tavignano Valley, another mountainous area is located in the north-eastern part of the island, which is somewhat lower than the main ridge. While the distance between the mountains and the east coast is partly up to 10 km, the west coast is rather steep with slopes starting right at the coast line.

During the measurement period from the middle of August to the end of October 2012, a large number of measurement devices were deployed on the Corsican Island complementary to permanently operated instruments. The distribution of the instruments enabled to measure differences in the atmospheric conditions over the island on the meso-γ scale. In Fig. 3.1 the measurement sites used in this thesis are displayed. The KITcube instruments were distributed over two main deployment sites, Corte (369 m MSL) and San Giuliano (39 m MSL). The Corte site was equipped to capture the atmospheric conditions and processes in the centre of the island and the San Giuliano site was set up to cover coastal to maritime atmospheric conditions. At both sites the instruments were installed on a flat, horizontally-homogeneous area of about 50 m x 50 m. The vegetation was mainly grass with some bushes. The instrumentation deployed at the two sites is listed in Tab. 3.1. Three mobile towers, which are part of the KITcube platform, were installed at Aléria (40 m MSL), Casaperta (52 m MSL) and Pont génois (183 m MSL) in the Tavignano Valley between the east coast and Corte. In addition to data derived from devices integrated in the KITcube platform by default, measurements from a microwave radiometer deployed at San Giuliano; from a permanent GPS network; from an automatic weather station (5 m MSL) and an operationally operated radiosonde system (5 m MSL) at Ajaccio; from the Dornier 128 aircraft based at Solenzara (14

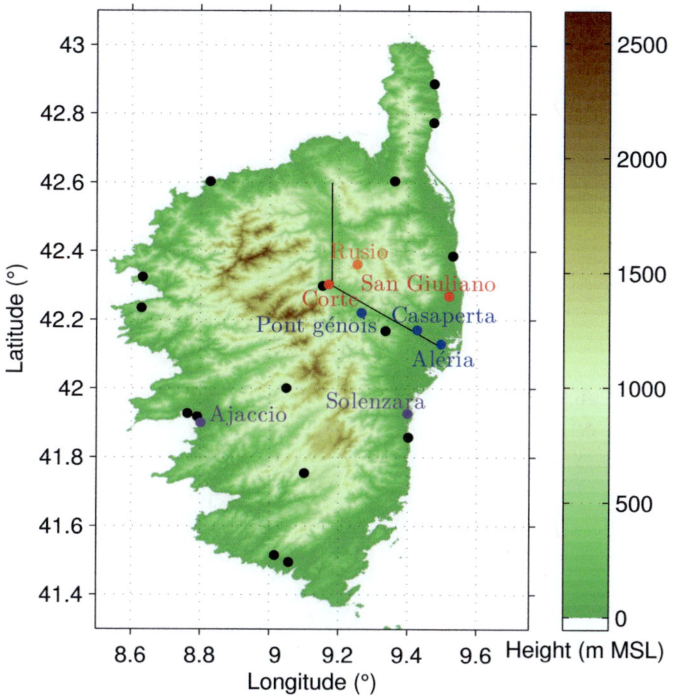

Figure 3.1.: Orography and measurement sites on the Corsican Island during the HyMeX field campaign: main deployment sites (red), mobile towers (blue), a temporarily installed GPS station (orange) and permanent GPS stations (black). The black lines indicate the axis of the Tavignano Valley. The displayed orography has a horizontal resolution of about 90 m.

m MSL); and from a temporarily installed GPS station at Rusio (1121 m MSL) were used in this thesis.

The case studies in this thesis are mainly based on data from Corte, which is roughly located in the middle of the Tavignano Valley. There, the valley exhibits a ridge-to-ridge distance in the order of 20 km. The orography surrounding Corte is rather complex and various characteristic ridge heights were determined for the different case studies depending on the respective flow conditions.

Table 3.1.: Observation systems deployed at the two main sites on the Corsican Island.

Corte	San Giuliano
Energy balance station	Energy balance station
2 flux stations	2 flux stations
20-m tower	
Scintillometer	
Radiosonde system	Radiosonde system
Microwave radiometer	Microwave radiometer
2 wind lidars	
Cloud radar	
Ceilometer	
Cloud camera	
Aerosol spectrometer	
	Rain radar

For the first case study (Sect. 5.1), characteristic ridge heights north-west and north-east of Corte were identified as the mean ridge heights plus their standard deviations. The terrain height was averaged for a band of 5 km width in the respective directions (Fig. 3.2a). North-west of Corte a mountain ridge rises up to 1540±360 m MSL (Fig. 3.2b). Behind the ridge in the north-west a valley occurs followed by another mountain ridge of roughly the same height. In the north-east a ridge rises to about 1270±230 m MSL (Fig. 3.2c). Thus, the characteristic ridge heights were 1900 m MSL (1530 m above Corte) in the north-west and 1500 m MSL (1130 m above Corte) in the north-east. Due to mainly zonal westerly flow conditions in the second and third case study (Sects. 5.2 and 5.3), the terrain height was averaged between 42.2° N and 42.4° N. (Fig. 3.3). The mean ridge height west of Corte was found to be about 1600±400 m MSL, which resulted in a characteristic ridge height of 2000 m MSL (1630 m above Corte).

3.2.2. Instrument Specifications

The KITcube platform consists of measurement systems, which provide information on the energy exchange at the surface, turbulence and mean conditions in the ABL and whole troposphere,

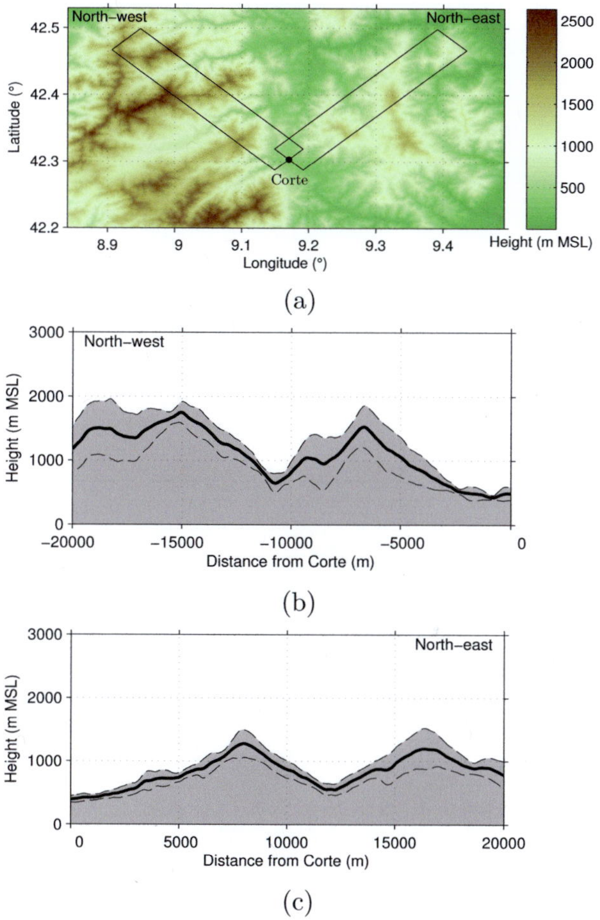

Figure 3.2.: Orography around Corte (a) and terrain height averaged for a 5-km wide band north-west (b) and north-east (c) of Corte. The averaged areas are marked by the black rectangulars in (a). In (b) and (c), the black thick line indicates the mean terrain height, the dashed lines the mean height plus/minus its standard deviation and the grey shading the characteristic terrain height.

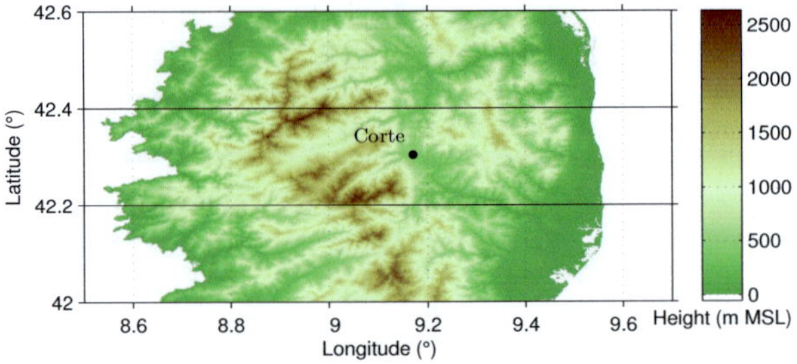

(a)

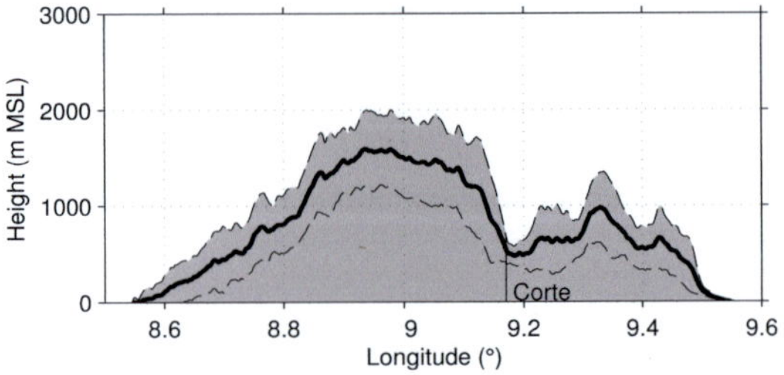

(b)

Figure 3.3.: Orography around Corte (a) and terrain height averaged between 42.2° and 42.4° N (b). The averaged area is marked by the black lines in (a). In (b), the black thick line indicates the mean terrain height, the dashed lines the mean height plus/minus its standard deviation and the grey shading the characteristic terrain height. The location of Corte is given by the vertical line.

respectively, and on cloud and precipitation properties. The systems of KITcube and additional instruments used in this thesis are described below. The complete instrumentation during the HyMeX field campaign is given by Kalthoff et al. (2013a).

Surface Energy Exchange and Near-Surface Observations

Energy-balance stations At Corte and San Giuliano, energy balance stations measured standard meteorological variables (temperature, humidity, wind speed and direction, air pressure, precipitation); radiation temperature of the surface; solar and reflected irradiance; and long-wave incoming and outgoing radiation (Kalthoff et al., 2006) with an integration time of 10 minutes. Sensible and latent heat fluxes; TKE and σ_w^2 were calculated for 30-minute intervals. The turbulent variables were measured with an ultrasonic anemometer/thermometer and a fast infrared hygrometer.

20-m tower At Corte, a 20-m-high meteorological tower provided wind speed, temperature, and humidity profiles up to 19 m AGL and wind direction at 20 m AGL with an integration time of 10 minutes. An ultrasonic anemometer/thermometer delivered wind speed and direction, TKE and σ_w^2 at 20 m AGL for 30-minute intervals.

Flux stations Two additional flux stations were installed at each main site. They delivered temperature, humidity and wind speed and direction with an integration time of 10 minutes and momentum and sensible heat fluxes, TKE and σ_w^2.

To gain information on the spatial variability of the meteorological variables at Corte and San Giuliano, temperature and humidity at 2 m AGL and wind speed, wind direction, TKE and σ_w^2 at 4 m AGL were averaged over available measurements from the energy-balance stations, the 20-m tower and the flux stations at each site. For post-processing of the turbulent variables, the eddy-covariance software package TK3.1 described by Mauder and Foken (2011) was used.

Scintillometer Additionally at Corte, a scintillometer designed by Scintec provided the spatially averaged sensible heat flux over homogeneous grass land along a laser beam of about 50 m length at 2 m AGL with an integration time of 1 minute. The beam was aligned along the valley axis.

Aerosol spectrometer At the same site, an aerosol spectrometer designed by Grimm delivered the particle concentration near the surface for particle sizes from 0.25 to 32 μm every minute.

Mobile towers Three battery-powered mobile towers were deployed in the Tavignano Valley (Fig. 3.1) to provide spatial information along the valley axis about temperature, humidity, wind speed and direction, air pressure, precipitation, TKE and σ_w^2. The standard meteorological variables were averaged over 5 minutes and TKE and σ_w^2 were calculated with the TK3.1 software package for 30-minute intervals.

Meteorological surface station as part of a GPS station As part of a GPS station, provided by GFZ Potsdam, near-surface temperature, wind speed and direction, humidity and air pressure were measured at Rusio with a 5 minute temporal resolution.

Automatic weather station An automatic weather station at Ajaccio operated by METEO FRANCE delivered temperature and humidity at 2 m AGL and wind speed and direction at 10 m AGL with a temporal resolution of 60 minutes.

Mean and turbulent atmospheric conditions

Radiosondes To obtain vertical profiles of temperature, humidity, wind speed, and wind direction in the troposphere, two radiosonde systems manufactured by Graw were operated at Corte and San Giuliano during Intense Observation Periods (IOPs). Depending on height, the 1 s raw data were averaged over 2 s (between ground and 3 km AGL), 4 s (between 3 and 10 km AGL), and 8 s ($>$ 10 km AGL). During IOPs, radiosondes were typically launched at 2- to 3-hour intervals. At Ajaccio, operational radiosoundings were carried out by METEO FRANCE twice a day, typically at 0100 and 1300 Local Time (LT$\widehat{=}$ Coordinated Universal Time (UTC) plus 2 hours). The sounding data are also used to derive significant levels, like LCL or LFC, and convection-related parameters, such as CIN, CAPE, the TT index and RH.

Microwave radiometer A ground-based scanning microwave radiometer by Radiometer Physics observed atmospheric emission at Corte. The instrument measures sky brightness temperature, T_B, at 14 frequencies, seven are distributed between 22.235 and 31.4 GHz along the wing of the 22.235 water-vapour line and seven between 51.26 and 58.0 GHz along the wing of the 60 GHz oxygen absorption complex. The technical specifications and calibration methods are described by Rose et al. (2005). The half power beam width is about 3.5° for the channels along the water-vapour line and 2.5° along the oxygen complex. A microwave radiometer designed by Radiometrics and provided by GFZ Potsdam was installed at San Giuliano. It measured T_B at 8 frequencies between 22.234 and 30 GHz along the wing of the water-vapour line.

Information on atmospheric variables from the measured T_B were obtained with retrievals provided by the University of Cologne. The retrieval algorithm is described by Löhnert and Crewell (2003) and Löhnert et al. (2009). It is based on a training set of radiosonde profiles measured at Ajaccio between 1990 and 2009. A one-dimensional emission-only forward integration of the radiative transfer equation is used to generate T_B at the radiometer frequencies from the radiosonde profiles and a random error is added to account for the error of real radiometer measurements.

Using linear regression and including quadratic terms, a relation between IWV and T_B is obtained, which can then be used to retrieve the IWV from T_B measured by radiometer. The IWV derived by microwave radiometers has in general a high accuracy of less than 1 kg m^{-2} (Pospichal and Crewell, 2007; Kneifel, 2008). To compare IWV derived from radiometer measurements with IWV calculated from radiosonde profiles, the radiometer IWV measured in the vertical stare mode was averaged over a 60-minute interval centred on the launchtime of the radiosonde. A correlation coefficient of 0.97 resulted at Corte and 0.91 at San Giuliano for the whole HyMeX campaign (Figs. 3.4a and b).

Furthermore, the vertically measured IWV was used to determine the water-vapour distribution between the individual radiosoundings. Firstly, the ratio of IWV measured by the microwave radiometer $IWV_{MWR}(t)$ to the linearly interpolated IWV from ra-

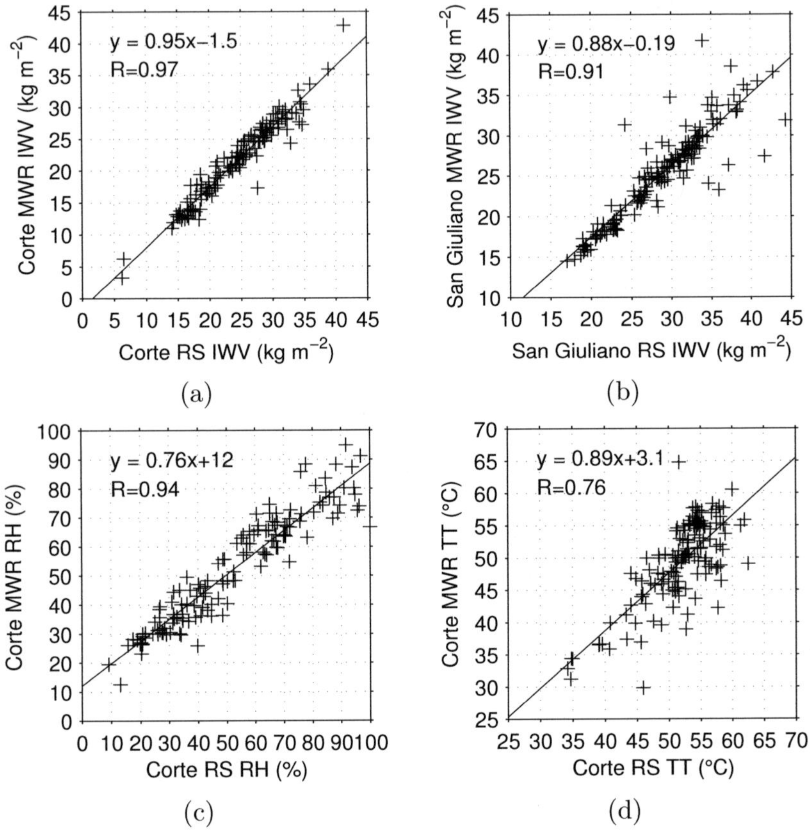

Figure 3.4.: IWV measured by radiometer (MWR) and radiosonde (RS) at Corte (a) and San Giuliano (b) and RH (c) and TT index (d) measured by radiometer and radiosondes at Corte during the whole HyMeX campaign. The radiometer values are averaged over a 60-minute interval centred on the launchtime of the radiosonde. The line of best fit is indicated by the straight line and its equation and the correlation coefficient are given.

diosonde ascents $\mathrm{IWV_{RS}}(t)$ was determined. To obtain the specific humidity distribution $q(z,t)$ (z: height AGL, t: time) the temporally linearly interpolated specific humidity field from radiosondes $q_{\mathrm{RS}}(z,t)$ was then multiplied by this ratio:

$$q(z,t) = q_{\mathrm{RS}}(z,t)\frac{\mathrm{IWV_{MWR}}(t)}{\mathrm{IWV_{RS}}(t)}. \tag{3.1}$$

The thus post-processed specific humidity profiles were used in this thesis. The water-vapour density distribution between the radiosoundings was estimated alike.

A scanning microwave radiometer is able to perform azimuth scans, i.e. scanning 360° at a fixed elevation angle, and elevation scans, i.e. scanning the upper half space at a fixed azimuth angle, and provides IWV along the line of sight. Azimuth scans above the horizon are found to be useful to detect spatial humidity variability of significant strength (Kneifel, 2008). The microwave radiometer at Corte regularly performed azimuth scans. The selected elevation angle of 19.8° allowed to scan above the orography surrounding the site. For comparison with vertical measured IWV, the values of the azimuth scans $\mathrm{IWV_{azimuth}}$ were normalized to a vertical column $\mathrm{IWV} = \mathrm{IWV_{azimuth}}\sin(19.8°)$ for each azimuth angle.

The opaque oxygen complex and the use of elevation scanning (Crewell and Lohnert, 2007) allows to obtain temperature profiles in 50 m steps close to the surface and in steps of 200 to 400 m at higher altitudes. Assuming horizontal homogeneity of the atmosphere within a horizontal radius of about 3 km around the radiometer, observed radiation systemtically originates from higher elevations the greater the elevation angle. Adding the information from elevation scans at 5.4°, 10.2°, 19.2°, 30.0°, and 42.0° for seven frequencies to the standard zenith observations and using linear regression the retrieval of temperature profiles with an accuracy of less than 1K and of inversions below about 1.5 km is possible (Crewell and Lohnert, 2007).

Humidity profiles derived from microwave radiometers have a very scarce resolution as only two of the seven available water-vapour channels are independent (e.g. Löhnert et al., 2009) and the humidity profiles show in all heights approximately the same temporal evolution. Thus, the gain of information obtained from these humidity profiles in addition to IWV is generally very low.

The obtain information on the pre-convective conditions on days when no radiosondes were available, RH and the TT index were calculated from radiometer measurements at Corte. These profiles were averaged over a 60-minute interval centred on the hour. Comparison of the parameters with the respective values derived from radiosoundings revealed a correlation coefficient of 0.94 for RH and 0.76 for the TT index (Figs. 3.4c and d). Because of the low vertical resolution of temperature and humidity profiles, CAPE and CIN calculated from the radiometer measurements did not correlate with CAPE and CIN obtained from radiosonde profiles and therefore were not used.

Infrared radiometer As part of the microwave radiometer at Corte, an upwards directed infrared radiometer measured the cloud-base temperature at wavelengths of 11.1 and 12 μm. In combination with temperature profiles, the cloud-base height could be derived.

Wind lidars Light detection and ranging (lidar) systems allow the detection of spatial atmospheric structures via active remote sensing. The basic principles of active remote sensing are described by e.g. Banta et al. (2013). The lidar technology is based on the transmission of electromagnetic radiation in a range between 200 nm and 10 μm. A monochromatic, narrow laser beam with a large coherence length is transmitted and the returning signal is detected as a function of time delay. The received power depends on the transmitted power, the scattering properties of the atmosphere, which is expressed via the backscatter cross section, attenuation losses during the two-way propagation of the signal through the atmosphere and the range from the instrument to the scattering target. The backscatter cross section describes the amount of energy scattered back towards the receiver at an angle of 180° from the direction of the transmitted signal. The total attenuation composes of absorption by gas molecules, hydrometeors and aerosol particles as well as scattering out of the beam in all directions, rather than backwards towards the receiver. Even without a transmitted signal, the receiver will detect a signal, the so-called background noise. The ratio between the standard deviation of the received signal strength, σ_S, and the standard deviation of the noise, σ_N, is described by the Signal-to-Noise Ratio

$$\mathrm{SNR} = 10\log_{10}\frac{\sigma_S}{\sigma_N} \tag{3.2}$$

with the unit dB. The detectability of the signal can be improved, i.e. the SNR can be increased, by averaging over more independent pulses, i.e. by increasing the integration time. Using SNR, instrument parameters and a constant absorption coefficient the backscatter intensity can be derived (e.g. Träumner, 2012).

Scattering targets for lidars are mainly aerosols and molecules depending on the wavelength of the transmitted beam, which allows measurements in the cloud-free atmosphere. On the other hand, the signal is attenuated by liquid water, which leads to a limited penetration of clouds or fog.

Wind lidars are designed to detect radial velocities via the doppler shift, i.e. the frequency difference between the transmitted and the returned signal, which occurs when the scattering target is moving. Thereby, the doppler effect appears twice, for the transmitted and the returned signal, and the doppler velocity describes the radial velocity of the scattering target. It has to be considered that variations of radial velocity within the pulse volume, which is specified by the range gate and the beam width, cannot be detected and the estimated radial velocity is an average over the whole pulse volume.

Two wind lidars were operated at Corte. One of the lidars is a pulsed 2-μm scanning wind lidar with a beam width of 75 mm at laser exit developed by Lockheed Martin, which delivers aerosol backscatter and radial velocity. A detailed description of the instrument is given by Träumner (2012). The average power of the system is 1 W, with a peak power of 4.5 kW. It is characterized by an effective pulse length, i.e. the minimum distance between two scattering targets that is necessary for the instrument to detect two separate entities, of 56 m and a pulse repetition frequency of 500 Hz. It further comes with a two-axis scanner, which allows to measure at all azimuth angles and down to an elevation angle of 0°, i.e. horizontal. To obtain information on the 3-dimensional wind field, the vertical stare mode (typical integration time of 1 s; measurement range from 375 to 5400 m AGL with a vertical range resolution of 46 m) was combined with azimuth and elevation scans. To prevent noise effects for each range gate, data with SNR

< -8 dB were typically excluded. In August 2012, the lidar was operated with an integration time of 0.1 s in the vertical stare mode. Because of very noisy data, the radial velocity was post-processed using a SNR threshold of -6 dB, removing outliers under the assumption that the radial velocity was normally distributed and averaging over 1 s. The uncorrelated noise of the vertical wind speed was estimated using a linear fit through the first lags of the auto-covariance function (Lenschow et al., 2000). The root of the difference between the interpolated value to lag zero and the calculated value of the auto-covariance function at this point gives the uncorrelated noise. For typical lidar settings, the uncorrelated noise was estimated to be less than 0.15 m s^{-1} (Träumner et al., 2011). As the wind lidar only supplies valid data when the aerosol content is sufficient, the layers with wind speed measurements reflect the distribution of the aerosol content. Azimuth scans were performed at elevation angles of 10° and 75° and elevation scans at azimuth angles of 0°, 45°, 90° and 135°. With a maximum rotational speed of 20° s^{-1}, the duration of two azimuth scans or four elevations scans amounted to about 2.5 minutes.

In order to cover the range between the surface and the lowest lidar measurement height, a second wind lidar with a wavelength of 1.54 μm manufactured by Leosphere was installed at Corte. It was available for parts of the HyMeX field campaign, starting at the end of September. Depending on the aerosol content, it provided data from 40 m AGL up to about 600 m AGL with an integration time of 7 s and a vertical range resolution of 20 m. As an indicator of data quality the carrier-to-noise ratio is used, which is equivalent to the SNR. To ensure good data quality, a carrier-to-noise ratio of -26 dB was applied. In normal operation mode, this small wind lidar has a fixed elevation angle of 75.2° and an azimuth scanner. By means of a 4-point stop and stare plan-position-indicator scan applying the velocity-azimuth display algorithm, the wind profile is calculated. During the HyMeX field campaign a special operation mode was run using a motorized tilted table, which allows for continuous vertical stare measurements and for the direct detection of vertical velocity.

Cloud radar In meteorology, radio detection and ranging (radar) systems are widely used to detect precipitation and cloud properties and their spatial distributions. The wavelength of the transmitted electromagnetic radiation is adjusted to the measurement purpose (e.g. Banta et al., 2013). Similar as for lidars, the received power depends on system properties, atmospheric attenuation and on the backscatter cross section. In the Rayleigh regime, where the target diameter, D_t, is much smaller than the transmitted wavelength, λ, the backscatter cross section can be written as (e.g. Banta et al., 2013)

$$\sigma_i = \pi^5 |K^2| \frac{D_t^6}{\lambda^4}, \tag{3.3}$$

where $|K^2|$ is related to the composition-dependent index of refraction. For a given target diameter, backscattering is inversely proportional to the fourth power of the wavelength, which means that shorter wavelengths are more efficiently scattered than longer wavelengths. For a fixed wavelength, the backscattered energy is proportional to the sixth power of target diameter. Because of that, backscattering in the Rayleigh regime is often dominated by a few large targets in the scattering volume. For example, rain radars transmitting signals with wavelengths of several centimetres are able to detect raindrops over large horizontal distances, as the attenuation is low. But on the other hand, cloud droplets are invisible to them, as they do not receive enough reflectivity. Cloud radars typically transmit signals with wavelengths less than 1 cm and are able to detect cloud droplets. Also in the absence of clouds, a cloud radar may detect clear-air returns (e.g. Banta et al., 2013; Kalthoff et al., 2013a), which are caused e.g. by insects or other floating material/debris, dust and Bragg scattering from refractive-index fluctuations. Via the doppler shift, the radial velocity is determined equivalent to the method for lidar systems.

Implemented in the KITcube platform, a scanning cloud radar with a frequency of 35.5 GHz (λ = 8.5 mm) manufactured by Metek provides radial velocity in clouds and partly in cloud-free conditions at Corte. A detailed instrument description is given by Grenzhäuser (2011). The cloud radar features an average power of 30 kW, a beam width of 0.7°, an effective pulse length of 30 m and a pulse repetition rate of 5 kHz. During the HyMeX campaign, the cloud radar was operated mainly in the vertical stare mode with an integration time of 10 s and a measurement range from about 150 to 15000 m AGL at a vertical resolution of 30 m. To remove noise, the data were smoothed using a running mean of 90 s, which still allowed to analyse convection-related structures in the mountain ABL.

Because the cloud radar requires a smaller scatterer concentration than the wind lidar to provide sufficient backscatter intensity and because the two instruments see different particles in cloud-free conditions, they provide data in various layers (Kalthoff et al., 2013a). But even when both instruments are operated in the vertical stare mode next to each other and return data in the same layer, they may observe different radial velocities, due to the different scattering targets and pulse volumes. Handwerker et al. (2008) reported that a cloud radar detects larger downward motions in clouds and in layers with rain compared to a wind lidar. The correlation coefficients between vertical velocity measurements with the wind lidar and cloud radar at 10 s resolution were 0.64 for positive and 0.70 for negative vertical velocities under cloud-free conditions on days with undisturbed mountain ABL evolutions (Fig. 3.5). Because the cloud radar overestimated negative and un-

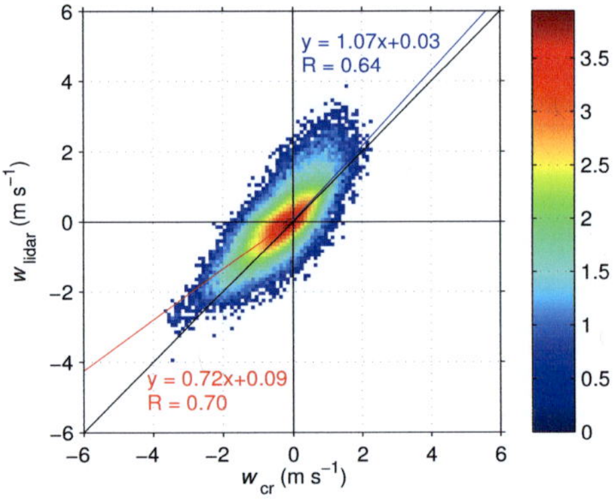

Figure 3.5.: Distribution of vertical velocities measured by wind lidar (index lidar) and cloud radar (index cr) under cloud-free conditions on days with undisturbed mountain ABL evolutions. For a more comprehensive image, colours indicating number density are scaled logarithmically, i.e. '2' corresponds to 10^2 values per $(0.1\ \mathrm{m\ s^{-1}})^2$. The lines of best fit for positive (blue line) and negative (red line) vertical velocities are indicated by the straight lines and their equations and the correlation coefficients are given.

derestimated positive vertical velocities, a correction function was applied to the vertical velocity measured by the cloud radar, w_{cr}. The corrected vertical velocity $w = aw_{\mathrm{cr}} + b$ ($a = 1.07$, $b = 0.03$ for $w_{\mathrm{cr}} \geqq 0$ and $a = 0.72$, $b = 0.09$ for $w_{\mathrm{cr}} < 0$) was used in this thesis.

Sodar Sonic detection and ranging (sodar) systems transmit a pulse of sound into the atmosphere, which scatters off turbulence-scale fluctuations in the acoustic index of refraction and returns a weak signal back to the receiver (e.g. Banta et al., 2013). The fluctuations that produce the backscattered signal consist of temperature gradients that occur over distances of half the transmitted wavelength and are mainly caused by turbulence. The acoustic backscatter signal thus depends on the variance of the vertical wind speed and the atmospheric stratification. Strong ambient wind may transport the transmitted signal downstream and may prevent the detection of a return signal. Typical transmitted frequencies range between 1 and 5 kHz, which correspond to wavelengths between about 10 and 30 centimetres. The vertical measurement range depends on the transmitted frequency, whereby lower frequencies enable higher measurement ranges.

The sodar system included in KITcube operates at multiple frequencies around 2 kHz, which allows a spatial resolution of 10 m and a measurement range between 30 and 1000 m AGL, although the maximum height with valid data strongly depends on the atmospheric conditions. The implemented sodar is a so-called phased-array sodar, which transmits acoustic signals in multiple directions. This allows to obtain the three-dimensional wind field profile. During the HyMeX campaign the sodar was positioned at San Giuliano and the wind profiles were calculated for 30-minute intervals.

Ceilometer A ceilometer is an aerosol lidar, which provides uncalibrated backscatter. As part of the KITcube platform, a ceilometer designed by Jenoptik was used to determine up to three different cloud-base heights between 150 and 15000 m AGL at Corte with an integration time of 1 minute. The system transmits laser pulses at 1.064 μm with a typical pulse energy of 8 μJ, an effective pulse length of 15 m and a pulse repetition frequency between 5 and 7 kHz.

Rain radar Volume filling information on precipitation was gathered by a mobile X-band radar manufactured by Gematronik positioned at San Giuliano. The radar is dopplerized and able to measure polarisation properties of the hydrometeors. The transmitted peak power is roughly 70 kW and the beamwidth is 1.3°. The radar reaches the total upper half-space down to an elevation of -2°. The horizontal range of the radar is about 100 km with a spatial resolution down to 75 m.

Cloud camera Visual information on cloud type and cloud cover at Corte was obtained by a camera with a fish-eye lens, which took pictures of the upper hemisphere every two minutes.

GPS network On Corsica, a permanent GPS network operated by Institut Geographique National (IGN) and ACTIPLAN company is installed existing of 16 GPS receivers. To obtain ZHD and T_{m}, surface pressure and temperature from the AROME WMED operational analysis were used (Olivier Bock, personal communication). The conversion of ZTD to IWV was performed by Olivier Bock (IGN LAREG) and Pierre Bosser (ENSG). All IWV data were obtained with the same algorithm (Bock et al., 2007) and were available with a temporal resolution of 60 minutes. The accuracy of IWV measured by GPS has been estimated to be 1 to 2 kg m^{-2} in midlatitudes, using intercomparisons with e.g. radiosondes or microwave radiometers (e.g. Rocken et al., 1995; Bock et al., 2005).

Aircraft On specific IOP days in September and October 2012, the Dornier 128 aircraft D-IBUF from TU Braunschweig was operated. It is equipped with in-situ sensors to measure meteorological variables, like temperature, humidity and the three-dimensional wind field and turbulence (Corsmeier et al., 2001) at a sampling rate of 100 Hz. At a mean ground speed of 60 m s^{-1} the horizontal resolution of the meteorological variables is less than 1 m. In this thesis, data from west-east flight tracks across the island at different levels were used to obtain information on the atmospheric conditions above the mountain ridge height.

4. Determination of the Convection-Layer and the Mountain ABL Height

The superposition of convection and mesoscale circulations over complex terrain impacts the atmospheric conditions and obviously causes differences in the CBL development and structure over complex compared to homogeneous terrain (Sect. 2.3). To analyse that, CBL heights over both terrain types were determined with different methods and compared (Sect. 4.1). The complex terrain was represented by the measurement site in the Tavignano Valley at Corte on the Corsican Island during HyMeX (Sect. 3). Within the HD(CP)^2 Observational Prototype Experiment (HOPE) the KITcube was deployed at Hambach (50.897° N, 6.464° E, 113 m MSL) near Jülich in western Germany in spring 2013. HOPE is part of the High Definition of Clouds and Precipitation for Climate Prediction (HD(CP)^2) initiative. The orographic elevations in the surrounding of the site Hambach are less than 150 m and the terrain was assumed to be approximately homogeneous, at least compared to Corte. During HOPE, the wind lidar was operated with a measurement range from 350 to 3100 m AGL at a vertical resolution of 25 m. For each terrain type, a day with a typical evolution of the atmospheric conditions in the absence of significant large-scale advection was chosen: 04 May 2013 for homogeneous terrain at Hambach and 24 August 2012 for complex terrain at Corte.

The mountain ABL characteristics apparently differ from those of the classical CBL over homogeneous terrain (Sect. 2.3). This suggests that the standard methods to determine the CBL height do not apply and an objective method based on humidity profiles was developed and implemented to detect the top of the mountain ABL (Sect. 4.2).

4.1. The Convection Layer

The implemented methods to determine z_i, overviewed in (Sect. 2.1.3), include the parcel method and the gradient method for potential temperature and specific humidity profiles measured by radiosondes and the gradient method for aerosol content profiles measured by wind lidar. In addition, a modified threshold method for σ_w^2 profiles calculated from wind lidar measurements was developed and applied.

Homogeneous Terrain

The profiles of potential temperature and specific humidity that developed in the afternoon at Hambach were typical for a CBL (Fig. 4.1a). An unstably stratified surface layer, a well-mixed layer above topped by an inversion could be distinguished. The specific humidity was about height-constant in the well-mixed layer decreasing with height around the inversion level. The CBL heights determined from radiosoundings with the gradient method for potential temperature ($z_i^{\Theta\mathrm{GM}} = 1105$ m AGL) and for specific humidity ($z_i^{q\mathrm{GM}} = 1055$ m AGL) resulted in very similar results. Because of a very high surface temperature, the parcel method yielded in $z_i^{\mathrm{PM}} = 1460$ m AGL.

To detect z_i from aerosol content profiles measured by the wind lidar, the gradient method was used. The largest negative peak of the first derivative of the backscatter intensity averaged for 10 s intervals was determined as instantaneous CBL heights (Träumner, 2012). Due to the averaging, gradients that arise from small-scale structures or noise are smothered to avoid erroneous heights. The so-detected heights are shown in Fig. 4.1d for an hour-long time period. The CBL height averaged over one hour, z_i^{Aerosol}, calculated to 1147±171 m AGL.

In addition, z_i was determined with a modified threshold method for σ_w^2 measured by wind lidar, hereafter named $z_i^{\sigma_w^2}$. The σ_w^2 profiles were calculated for 60-minute intervals from the post-processed wind speed data (Sect. 3.2.2) with an integration time of 1 s. When the averaging interval contained less than 60% of valid data, the whole range gate was excluded from further analysis. Additionally, a linear trend was removed from the vertical wind speed time series. The used threshold value of $\sigma_{w,\mathrm{th}}^2 = 0.25$ m^2

s^{-2} was higher than the values used in previous studies (Tucker et al., 2009; Träumner et al., 2011) to account for generally higher turbulence and σ_w^2 values above the CBL over complex terrain (Serafin and Zardi, 2010b).

To ensure that the turbulence in the CBL was surface-based and buoyancy-driven, an additional criterion was defined using the skewness of the vertical wind speed

$$S = \frac{\overline{w'w'^2}}{\sigma_w^3}. \qquad (4.1)$$

Like σ_w^2, it was calculated from the linearly detrended vertical wind speed for 60-minute intervals provided that the period contained more than 60 % of valid data. The vertical transport of the instantaneous variance w'^2 by turbulence itself is described by $\overline{w'w'^2}$. In the CBL, the dominant source term for w'^2 is buoyancy generation, which is proportional to the sensible heat flux (Garratt, 1994). As it is maximum at the surface and decreases with height, there is an upward transport of w'^2 and S is positive throughout the CBL (LeMone, 1990; Hogan et al., 2009). Therefore, $z_i^{\sigma_w^2}$ was determined as the range gate above which (i) σ_w^2 dropped below $\sigma_{w,\mathrm{th}}^2$ and (ii) the great majority of S (>60%) was positive below this level. If σ_w^2 was larger than $\sigma_{w,\mathrm{th}}^2$ throughout all range gates with valid data, the highest of these range gates was used as $z_i^{\sigma_w^2}$ given that criterion (ii) still applied. This implies that convection reached up to the top of the layer with a high aerosol content. The systematic and random errors after Lenschow et al. (1994) can be used to estimate the uncertainty of the calculated σ_w^2 and S. The systematic error results from the usage of time averages instead of ensembles averages and the random error follows from discrepancies from the true value. Typically, the systematic error is much lower than the random error. The total error was calculated as the sum of the systematic and random errors. By subtracting and adding the total errors to the σ_w^2 and S profiles, modified profiles were received, for which additional $z_i^{\sigma_w^2}$ were calculated. The spread of the three $z_i^{\sigma_w^2}$ was interpreted as uncertainty range of the CBL height. An example of $z_i^{\sigma_w^2}$ determined in this way for the period between 1600 and 1700 LT is shown in Figs. 4.1b and c. Associated with strong up- and downdraughts, σ_w^2 values were rather high up to 1180 m AGL. Adding and subtracting the total errors resulted in an uncertainty for $z_i^{\sigma_w^2}$ of +25 m and -75 m. S was positive throughout the CBL, decreasing towards $z_i^{\sigma_w^2}$.

Because the profiles of potential temperature and specific humidity measured by radiosondes were a result of previous turbulent mixing, $z_i^{\sigma_w^2}$ and z_i^{Aerosol} were stored at the end of the time intervals and compared with $z_i^{\sigma_w^2}$, $z_i^{\Theta\mathrm{GM}}$ and $z_i^{q\mathrm{GM}}$ from subsequent radiosonde profiles; i.e. z_i from radiosondes launched at 1700 LT were compared with z_i from σ_w^2 or aerosol content profiles calculated for the time interval between 1600 and 1700 LT.

Over homogeneous terrain, the evolution of z_i determined with the different methods agreed fairly well (Fig. 4.2a). All methods show a roughly linear increase of z_i during the morning. The maximum z_i values were about 1500 m AGL detected with the parcel method; about 1200 m AGL with the gradient method for potential temperature and for aerosol content and from wind speed variance and skewness profiles; and about 1000 m AGL with the gradient method for specific humidity. This indicates that turbulent mixing was the dominating process in the CBL and the vertical distributions of potential temperature, specific humidity and aerosols were mainly determined by convection. Due to the

Figure 4.1. ***(facing page)***: Potential temperature and specific humidity measured by radiosonde at 1700 LT (a); variance, σ_w^2, and skewness, S, of vertical wind speed measured by wind lidar averaged between 1600 and 1700 LT (b); and vertical wind speed (colour coded) (c) and backscatter (colour coded) with instantaneous CBL heights (dots) (d) measured by wind lidar. All measurements are taken at Hambach on 04 May 2013. In (a), z_i calculated with the parcel method (z_i^{PM}) and with the gradient method for potential temperature ($z_i^{\Theta\mathrm{GM}}$) and specific humidity ($z_i^{q\mathrm{GM}}$) are shown. In (b), the error bars show the total error and the vertical dashed black line indicates the used variance threshold and the zero skewness line, respectively. In (b), (c) and (d), the solid line marks z_i calculated from σ_w^2 profiles ($z_i^{\sigma_w^2}$ in (b) and (c)) and from aerosol content profiles (z_i^{Aerosol} in (d)) and the dashed line marks the lower and the crosses the upper boundaries of the respective uncertainty range.

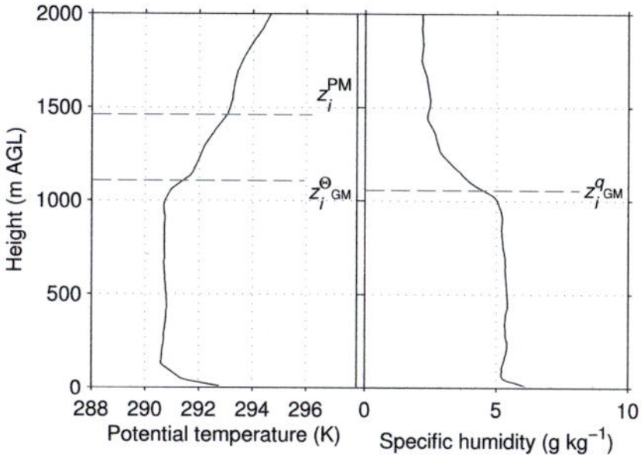

2000
1500
1000
500
0
Height (m AGL)
z_i^{PM}
$z_i^{\Theta}{}_{GM}$
$z_i^{q}{}_{GM}$
288 290 292 294 296
Potential temperature (K)
0 5 10
Specific humidity (g kg^{-1})

(a)

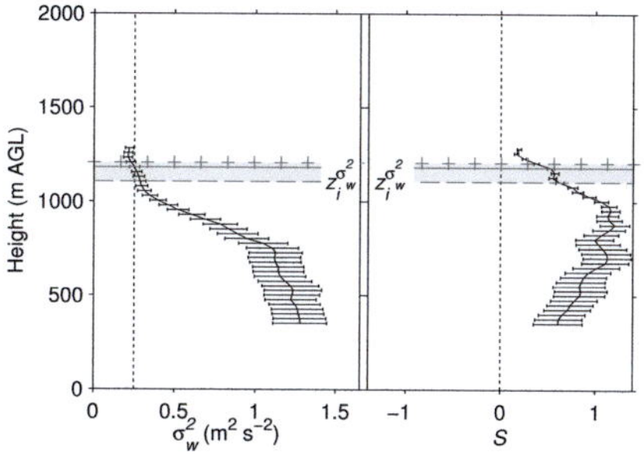

2000
1500
1000
500
0
Height (m AGL)
$z_i^{\sigma_w^2}$
0 0.5 1 1.5
σ_w^2 (m^2 s^{-2})
−1 0 1
S

(b)

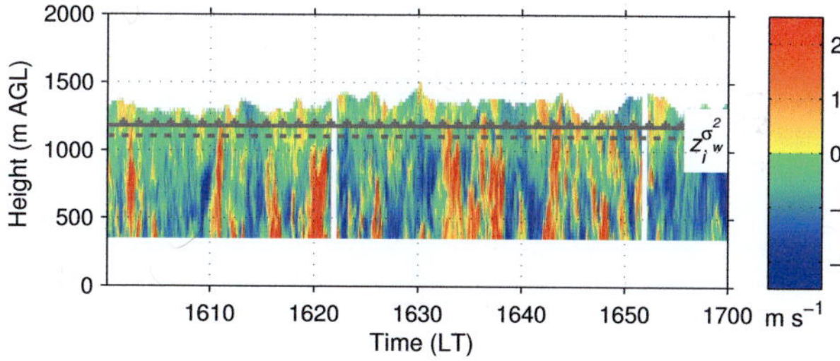

2000
1500
1000
500
0
Height (m AGL)
$z_i^{\sigma_w^2}$
1610 1620 1630 1640 1650 1700
Time (LT)
2
1
0
−1
−2
m s^{-1}

(c)

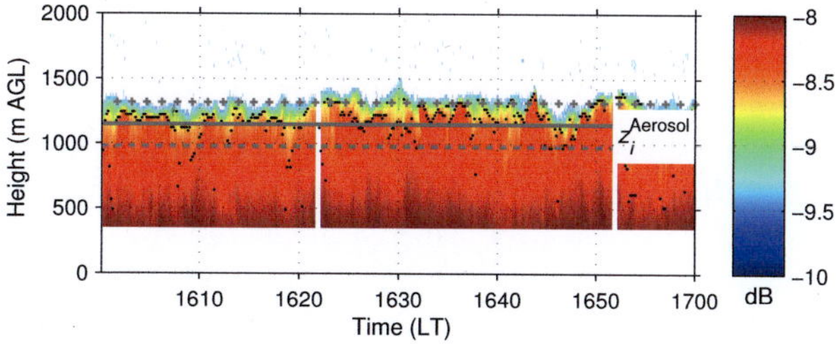

2000
1500
1000
500
0
Height (m AGL)
$z_i^{Aerosol}$
1610 1620 1630 1640 1650 1700
Time (LT)
−8
−8.5
−9
−9.5
−10
dB

(d)

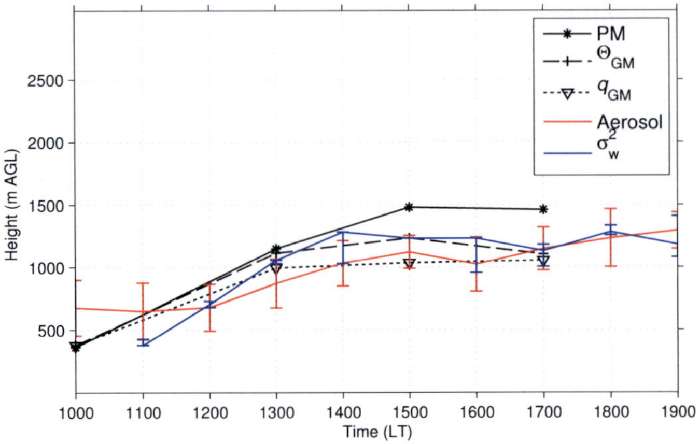

(a)

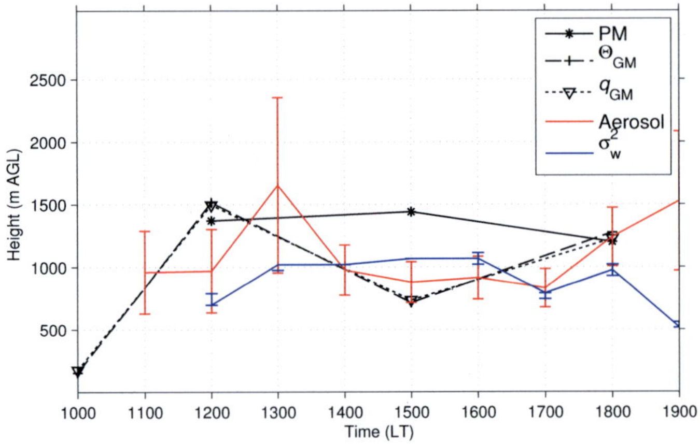

(b)

Figure 4.2.: Time series of z_i determined from radiosonde profiles with the parcel method (PM) and with the gradient method for potential temperature (Θ_{GM}) and specific humidity (q_{GM}); and from aerosol content profiles (Aerosol) and from wind speed variance and skewness profiles (σ_w^2) measured by wind lidar over homogeneous terrain at Hambach on 04 May 2013 (a) and over complex terrain at Corte on 24 August 2012 (b).

lowest measurement range gate of the wind lidar at around 350 m AGL, no growth of convection and thus no increase of $z_i^{\sigma_w^2}$ could be observed in the early morning. Before 1200 LT, z_i^{Aerosol} did not vary significantly, as it probably reflected the height of the residual layer at about 700 m AGL, which was associated with a high aerosol content. Consequently, the top of the aerosol layer was not a good indicator for z_i during this period.

Complex Terrain

In contrast, the various methods to determine z_i often produced significantly different heights over complex terrain (Fig. 4.2b). The temporal evolution of z_i determined from the profiles of mean variables did not show the typical diurnal cycle as it did over homogeneous terrain, i.e. a strong increase in the morning and approximately constant heights in the afternoon. Between 1000 and 1200 LT, $z_i^{\Theta_{\mathrm{GM}}}$ and $z_i^{q_{\mathrm{GM}}}$ increased by more than 1300 m, followed by a decrease of about 800 m until 1500 LT and another weaker increase until 1800 LT. Applying the parcel method, z_i^{PM} varied between about 1200 and 1440 m AGL during the afternoon. z_i^{Aerosol} ranged mainly around 1000 m AGL, only at 1300 LT it was as high as 1700 m AGL. Its uncertainty was often much larger than over homogeneous terrain. Calculated from σ_w^2 and S profiles, $z_i^{\sigma_w^2}$ increased from 700 m AGL at 1200 LT to 1070 m AGL at 1600 LT. Before 1200 LT the measurement range of the wind lidar inhibited the detection of shallow z_i.

The differences in z_i over complex terrain are illustrated in detail using the example of 1200 LT. The height detected with the parcel method z_i^{PM} was about 1400 m AGL (Fig. 4.3a). The profiles of potential temperature and specific humidity were not constant with height below that level and the gradients detected as $z_i^{\Theta_{\mathrm{GM}}}$ and $z_i^{q_{\mathrm{GM}}}$ were rather weak. The σ_w^2 profile averaged between 1100 and 1200 LT revealed turbulence only up to about 800 m AGL, due to strong up- and downdraughts (Fig. 4.3b and c). Around $z_i^{\sigma_w^2}$ the aerosol content decreased with height (Fig. 4.3d). However, because this gradient was often weak and the aerosol content was rather high up to about 1700 m AGL, the instantaneous heights varied a lot resulting in a 60-minute mean of $z_i^{\mathrm{Aerosol}} = 970 \pm 330$ m AGL. This indicated that other processes than turbulent mixing significantly contributed to the evolution of the potential temperature, specific humidity and aerosol content distribution over complex terrain. The profiles of mean atmospheric variables differed significantly from the ideal structure of a classical CBL (Sect. 2.1 and Fig. 4.1). Therefore, from the existing measurements it was not clear which height was the 'true' z_i. To solve this problem, the purpose of the z_i determination was recalled. The goal was to determine the CBL, i.e. the layer affected by surface-based, buoyancy-driven turbulent mixing. Because σ_w^2 is the only available variable that emphasizes the process and not the result of turbulence, it was reasonable to use σ_w^2 profiles to detect the height of this layer. Since this layer did not show the structure of a classical CBL, it was termed convection layer instead of CBL to avoid confusion.

When additional vertical wind speed measurements with the second wind lidar in the layer below about 400 m AGL were available, shallow convection layers and the convection-layer growth in the morning could be detected. This was e.g. the case at Corte on 02 October 2012 (Sect. 5.2). Thereby, σ_w^2 was calculated for combined vertical wind speed profiles from the two wind lidars with 10 s resolution, because of the higher integration time of the second wind lidar. The procedure agreed with the one described above.

Figure 4.3. *(facing page)*: Potential temperature and specific humidity measured by radiosonde at 1200 LT (a); variance, σ_w^2, and skewness, S, of vertical wind speed measured by wind lidar averaged between 1100 and 1200 LT (b); and vertical wind speed (colour coded) (c) and backscatter (colour coded) with instantaneous CBL heights (dots) (d) measured by wind lidar. All measurements are taken at Corte on 24 August 2012. In (a), CBL heights calculated with the parcel method (z_i^{PM}) and with the gradient method for potential temperature ($z_i^{\Theta\mathrm{GM}}$) and specific humidity ($z_i^{q\mathrm{GM}}$) are shown. In (b), the error bars show the total error and the vertical dashed black line indicates the used variance threshold and the zero skewness line, respectively. In (b), (c) and (d), the solid line marks z_i calculated from σ_w^2 profiles ($z_i^{\sigma_w^2}$ in (b) and (c)) and from aerosol content profiles (z_i^{Aerosol} in (d)) and the dashed line marks the lower and the crosses the upper boundaries of the respective uncertainty range.

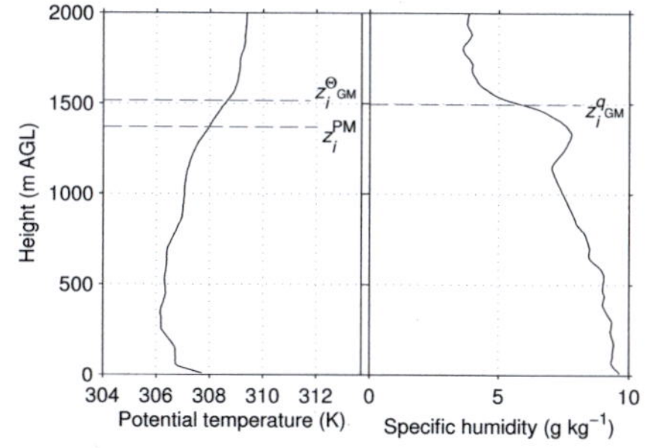
Height (m AGL)
Potential temperature (K)
Specific humidity (g kg−1)
z_i^{Θ} GM
z_i^{PM}
z_i^{q} GM

(a)

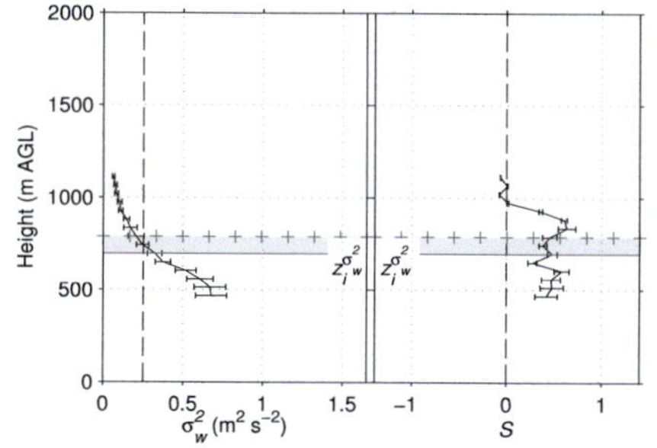
Height (m AGL)
σ_w^2 (m2 s−2)
S
$z_i^{\sigma_w^2}$

(b)

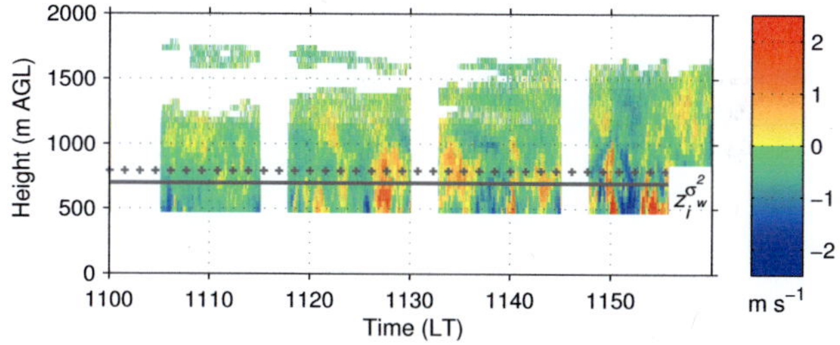
Height (m AGL)
Time (LT)
$z_i^{\sigma_w^2}$
m s−1

(c)

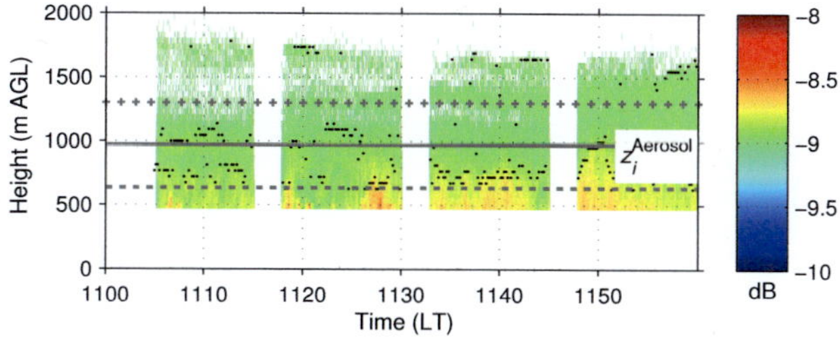
Height (m AGL)
Time (LT)
$z_i^{Aerosol}$
dB

(d)

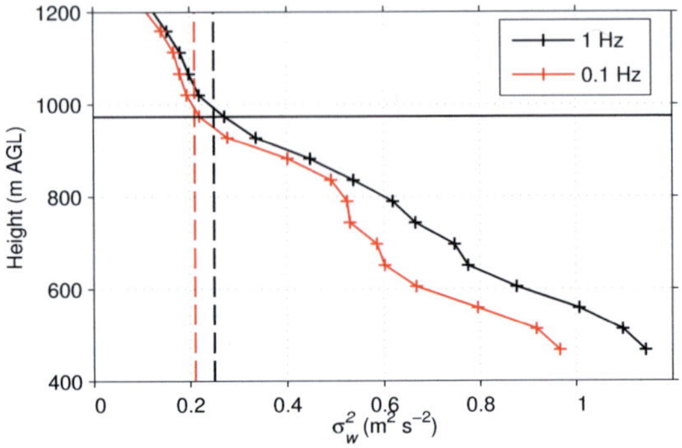

Figure 4.4.: Variance σ_w^2 of vertical wind speed measured with a wind lidar at Corte at 1800 LT on 24 August 2012. The vertical dashed lines indicate the used variance thresholds and the horizontal line marks the height of the convection layer.

Due to the cut-off of the higher frequencies in the σ_w^2-spectrum, σ_w^2 calculated from 0.1 Hz data was in average about 0.03 m^2 s^{-2} lower than σ_w^2 calculated from 1 Hz data. The comparison was done for vertical wind speed measured by the first wind lidar and averaged over the days with undisturbed mountain ABL evolutions at Corte, listed in Tab. B.1 in the appendix. To assure that the detected convection-layer heights remained the same, the threshold was reduced to $\sigma_{w,\mathrm{th}}^2 = 0.21$ for the 0.1 Hz data. The need to apply this reduction is illustrated by the example in Fig. 4.4. Using a threshold of $\sigma_{w,\mathrm{th}}^2 = 0.25$ for the 0.1 Hz data would have resulted in a lower convection-layer height by 46 m, i.e. one range gate. Applying the new threshold, the convection-layer heights determined with 1 Hz and 0.1 Hz data agreed in 70 % and differed by not more than one range gate in 95 % of all cases. In contrast, with the original threshold only 42 % of the convection-layer heights determined with the different temporal resolution would agree and 25 % would differ by more than one range gate.

4.2. The Mountain ABL

In the absence of large-scale advection and phase changes a change of humidity in an atmospheric layer may be caused by horizontal or vertical advection at the mesoscale or turbulent transport of water vapour (Sect. 2.5). An increase of humidity in an elevated layer normally indicates that the air has originated from a layer adjacent to the Earth's surface, where water vapour is added to the atmosphere due to evapotranspiration. Over homogeneous terrain at Hambach, the specific humidity profiles measured by radiosondes (Fig. 4.5a) reflect the CBL heights varying between 1000 and 1500 m AGL in the afternoon (Fig. 4.2a). The variations between the profiles at 1000 LT and 1300 LT were probably caused by large-scale processes.

In contrast, the profiles of specific humidity measured over complex terrain at Corte showed strong diurnal variations up to about 3500 m AGL (Fig. 4.5b), i.e. none of the methods used to determine z_i returned the top of the layer with diurnal variations as z_i (Fig. 4.2b). The layer top was around 2500 m above the convection-layer height determined as $z_i^{\sigma_w^2}$. In the morning, the humidity increased in elevated layers of several hundred metres of depth, e.g. at around 2000 m AGL at 1000 LT and at around 2500 m AGL at 1200 LT (Fig. 4.5b). During the afternoon, the moist layer extended over a layer much deeper than 1000 m, e.g. at 1500 LT humidity between about 1700 and 3500 m AGL was up to 4.5 g kg^{-1} higher than in the same layer in the morning. At 1800 LT, the humidity was still high below about 2800 m AGL. Until 2100 LT, the humidity above about 1400 m AGL decreased to approximately the same values as in the morning. The changes of humidity in about the lower 1000 m AGL were caused by the evolution of the convection layer as well as by the impact of the valley-wind system in the Tavignano Valley, which will be described in the next section. In contrast, the humidity changes in the elevated layers could not result from the convection-layer growth in the valley or the valley-wind system, but were probably caused by other mesoscale processes, like advective venting.

To determine the top of the mountain ABL, the height up to which a temporal increase of humidity occurred was detected. For this purpose, the difference between subsequent profiles of specific humidity Δq measured by radiosondes was calculated and the elevated layer with the maximum humidity increase Δq_{max} was detected. The top of the mountain ABL was then defined as the layer where the specific humidity difference above the layer of Δq_{max} dropped below $\Delta q_{\text{max}}\, \mathrm{e}^{-1}$, with e being Euler's number. In

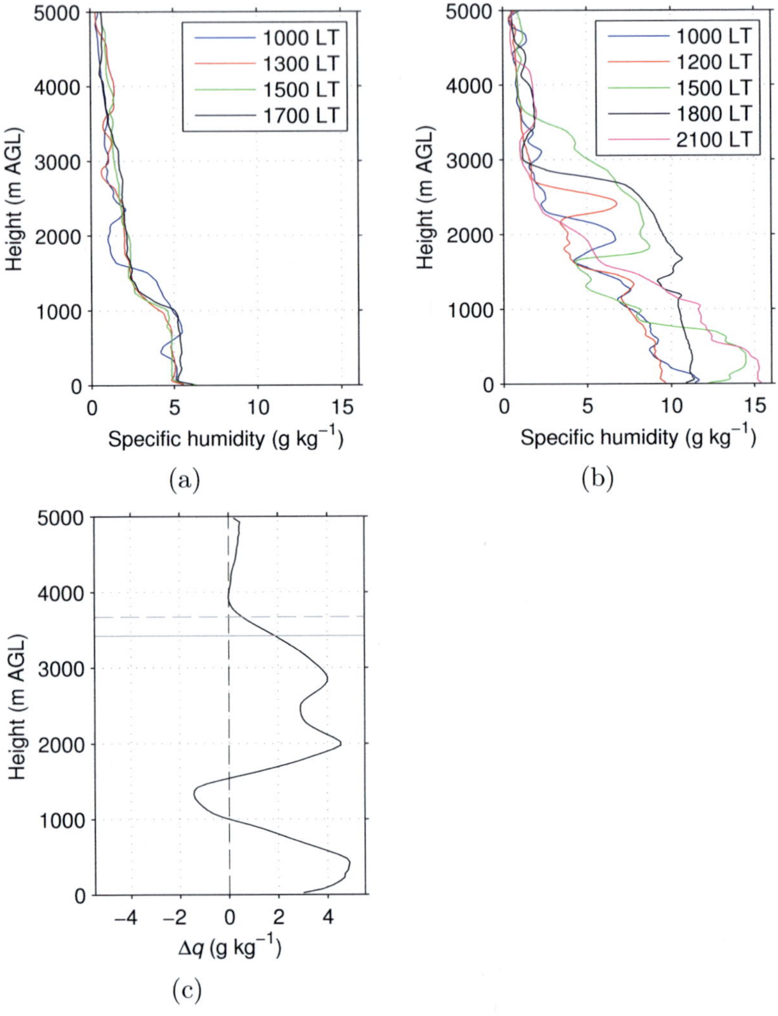

Figure 4.5.: Specific humidity measured by radiosondes over homogeneous terrain at Hambach on 04 May 2013 (a) and over complex terrain at Corte on 24 August 2012 (b); and specific humidity change Δq from 1200 to 1500 LT at Corte on 24 August 2012 (c). In (c), the solid and dashed grey lines indicate the mountain ABL heights determined with the Euler's number criterion and the 10 %-criterion, respectively.

addition, the layer was determined, where Δq dropped below 10 % of Δq_{max}. Because these methods were based on the humidity increase, they were only suitable to detect the growth of a deep mountain ABL, while a decrease of the mountain ABL height or its cessation could not be determined.

Figure 4.5c illustrates the detection of the mountain ABL in the Tavignano Valley using the radiosoundings at 1200 and 1500 LT at Corte. The maximum elevated humidity change occurred at around 2000 m AGL. The layer where Δq was significantly lower than Δq_{max} was detected at 3420 m AGL using the e^{-1}-criterion and at 3670 m AGL with the 10 %-criterion. Both heights differed by less than 10 % and in the following, the Euler's number criterion has been applied to detect the top of the mountain ABL. For comparison, the convection-layer height at 1500 LT amounted to 1060 m AGL (Fig. 4.2b), i.e. it was about 2500 m lower than the mountain ABL height.

5. The Evolution of the Mountain ABL

From the HyMeX measurement period, three days were selected, when the evolution of the mountain ABL at Corte differed completely. Because the mountain ABL represented the atmospheric layer that underwent a diurnal variation due to convection and thermally driven circulations, the upvalley-wind system in the Tavignano Valley was used as a criterion to distinguish between the different cases. It was assumed that the mountain ABL evolution was related to the evolution of the valley-wind system, characterized by the onset of upvalley wind in the morning and of downvalley wind in the evening and a constant upvalley-wind direction in between. Upvalley wind at the measurement station at Corte is associated with a south-easterly flow and downvalley wind with a north-westerly flow. This agrees with the orientation

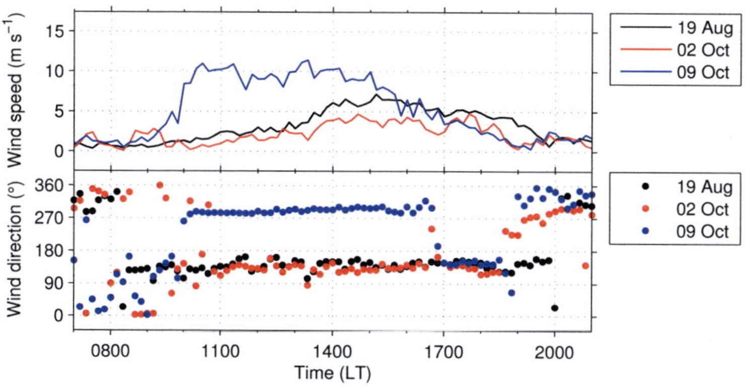

Figure 5.1.: Horizontal wind speed at 19 m AGL and wind direction at 20 m AGL at Corte on 19 August, 02 October and 09 October 2012.

of the axis of the Tavignano Valley between Corte and the east coast (Fig. 3.1).

Figure 5.1 shows wind speed and direction at Corte for the selected days: on 19 August 2012, an uninterrupted upvalley wind developed lasting from about 0830 to 2000 LT (Sect. 5.1). An upvalley wind existed for most of the day on 02 October 2012, although there was a short-term interruption in the afternoon between about 1600 and 1700 LT (Sect. 5.2). In contrast, no upvalley wind at all developed at Corte for most of the day on 09 October 2012 (Sect. 5.3), although a clear sky and high surface sensible heat flux would have enabled the evolution of convection and thermally driven circulations.

For each day, the processes and conditions that allowed, interrupted or inhibited the evolution of a mountain ABL were analysed in detail. In the following section only the day and month are given in the date information, because the analysis was exclusively based on data from the HyMeX field campaign in 2012.

5.1. The Undisturbed Mountain ABL Evolution: Case Study of 19 August 2012

5.1.1. Characteristics of the Mountain ABL at Corte

On 19 August, the Corsican Island was under the influence of a stationary ridge in the mid-troposphere with a north-south-oriented axis stretching across Central Europe and a high-pressure system below. Weak north-easterly large-scale wind was present in the free atmosphere above the island (Fig. 5.2).

During the day, a complex flow structure with three distinct wind layers evolved in the Tavignano Valley at Corte. The single layers are shown in Fig. 5.2. The upvalley-wind layer was defined as the layer with wind directions between 90° and 180° adjacent to the surface. An upvalley wind with maximum wind speeds of 7 m s^{-1} prevailed in the Tavignano Valley between about 1000 and 2000 LT (Fig. 5.3), reaching depths of 500 m at noon and 1000 m in the afternoon (Fig. 5.2). The onset of the upvalley wind at Corte was identified by an increase in near-surface wind speed to

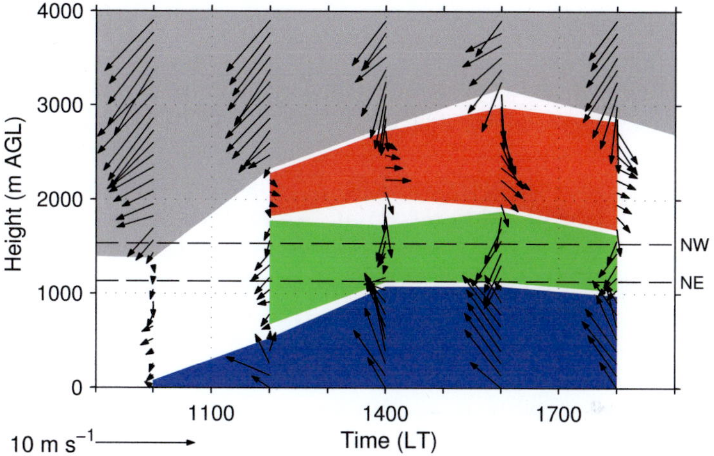

Figure 5.2.: Horizontal wind vector (arrows) measured by radiosondes at Corte on 19 August. The different wind layers are indicated by the shading: upvalley-wind layer (blue), wind layer I (green) and wind layer II (red). The grey shading marks the free atmosphere. The characteristic ridge heights north-west (NW) and north-east (NE) of Corte are indicated by the black dashed lines.

about 2 m s^{-1} rather than by a change of wind direction, which occurred earlier (Fig. 5.3) and was due to local processes. Above the upvalley-wind layer, two layers with different wind regimes existed between 1200 and 1800 LT (Fig. 5.2): wind layer I was characterized by wind directions between 0° and 90° and wind layer II was defined as the layer above wind layer I with wind directions between 270° and 360°. Above these wind layers, the large-scale flow characterized by wind directions between 0° and 90° persisted. The vertical extent of the different wind layers seemed to be related to the characteristic ridge heights north-west and north-east of Corte. The upvalley-wind layer extended to the surrounding ridges, i.e. it filled the whole valley in the vertical. Wind layer I mostly occurred above the north-easterly ridge and reached up to about 1800 m AGL, while wind layer II started a little higher than the north-westerly ridge and extended to about 3000 m AGL. That means, the evolution of three layers with different wind regimes disturbed the large-scale flow up to about 3400 m MSL.

Parallel to the evolution of the different wind layers, the specific humidity changed (Fig. 5.4a). The specific humidity was constant with height up to about 1200 m AGL at 1200 LT. The high temporal resolution of the distribution revealed that the specific humidity increased between 1000 and 1200 LT within a large part of wind layer I and decreased in the upvalley-wind layer below (Fig. 5.4a). As the IWV at Corte remained approximately constant during this time (Fig. 5.3), a redistribution of humidity obviously occurred between these layers. After around 1230 LT, the specific humidity in the upvalley-wind layer increased (Fig. 5.4a). Subsequently, the specific humidity assumed an increasingly height-constant profile up to about 1800 m AGL. Within a large part of wind layer II, the specific humidity increased in the course of the day. The increase of humidity in the various layers contributed to the IWV increase after around 1230 LT (Fig. 5.3). The re-onset of north-easterly large-scale wind above about 2200 m AGL between 1800 and 2100 LT was accompanied by a specific humidity decrease in the respective layer (Fig. 5.4a). As the mountain ABL top was determined by the increase of specific humidity, a mountain ABL height ranging between about 1500 m AGL at 1200 LT and 2800 m AGL at 1800 LT was detected (Fig. 5.4a), i.e. the maximum height of the mountain ABL approximately agreed with the height of the maximum vertical extent of wind layer II (Fig. 5.2).

A wind lidar and cloud radar measured the vertical wind speed at Corte (Figs. 5.4b and c). Between about 1130 and 1330 LT, both instruments detected very strong subsidence with embedded convective cells below about 1200 m AGL, i.e. within the whole valley in vertical direction. One-hour mean values measured by the wind lidar were as low as -1 m s^{-1}. The mean strong subsidence was observed by two independent remote sensing systems and also affected the ascent rate of a radiosonde at 1200 LT. In the same layers where the wind lidar and cloud radar detected mean subsidence, the ascent rate was about 1 m s^{-1} lower than the mean ascent rate (Fig. 5.5).

During the period with subsidence, the convection layer grew from about 600 m AGL at 1200 LT to about 1200 m AGL at 1300 and 1400 LT (Fig. 5.4b). Before 1200 and after 1700 LT, the convection-layer height was below the lowest detection level of the

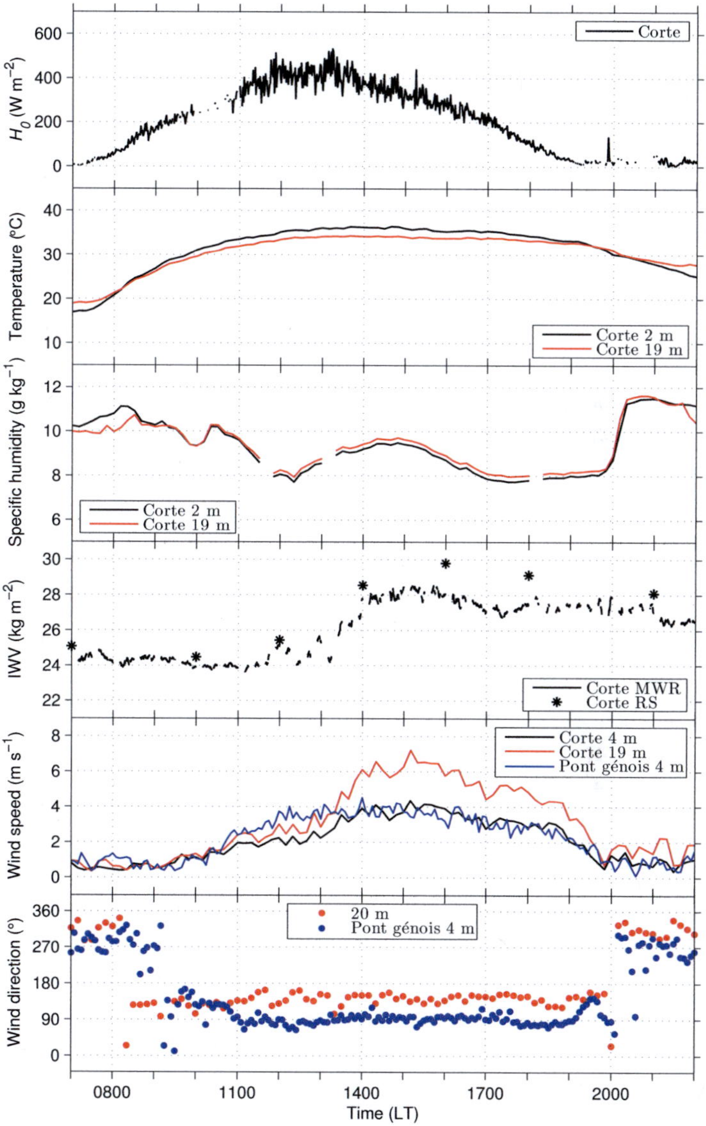

Figure 5.3.: Surface sensible heat flux, H_0, measured by scintillometer; temperature; specific humidity; and IWV at Corte and horizontal wind speed and direction at Corte and Pont génois on 19 August. IWV was measured by a microwave radiometer and calculated from radiosoundings.

wind lidar. The benefit of using the skewness as an additional criterion when detecting the convection-layer height from vertical wind-speed data (Sect. 4.1), became evident for the convection-layer height at 1200 LT. In this case, a convection-layer height could not be determined using the original profiles and the profiles with the added error due to elevated turbulence (Fig. 5.6a). Only for the profiles with the subtracted error was a convection-layer height of around 600 m AGL detected. The σ_w^2 profiles did not clearly resemble the convection-layer structure because of the elevated turbulence. In contrast, in the S profile it could explicitly be distinguished between surface-based turbulence (positive S) and elevated turbulence (negative S). The difference between the two turbulence types was also evident in vertical wind speed measurements by wind lidar and cloud radar (Figs. 5.4b and c). At 1300 and 1400 LT, no uncertainty range of the convection-layer height could be detected, because σ_w^2 did not fall below $\sigma_{w,\mathrm{th}}^2$ in any of the profiles (Fig. 5.4b). Hence, the highest range gate with valid wind-lidar data was determined to be the convection-layer height in all profiles.

Above the layer with subsidence elevated vertical motions were observed by the cloud radar and also by the wind lidar provided that enough scattering targets existed (Figs. 5.4b and c). Due to

---

Figure 5.4. ***(facing page)*:** Specific humidity (colour-coded), potential temperature (black isolines) and horizontal wind vector (arrows) measured by radiosondes (a); vertical wind speed measured by a wind lidar (b); and vertical wind speed measured by a cloud radar (colour-coded) and specific humidity measured by radiosondes (black isolines) (c) at Corte on 19 August. Specific humidity is interpolated using microwave radiometer and radiosonde data and potential temperature is temporally linearly interpolated between the radiosoundings. The mountain ABL height is indicated by the thick black dashed line. The characteristic ridge heights north-west (NW) and south-west (SE) are indicated by the thin black dashed lines. In (b), the solid horizontal lines mark the convection-layer height and the thick dashed horizontal lines mark the lower and the crosses mark the upper boundaries of the uncertainty range, if present. Note the different temporal and vertical ranges.

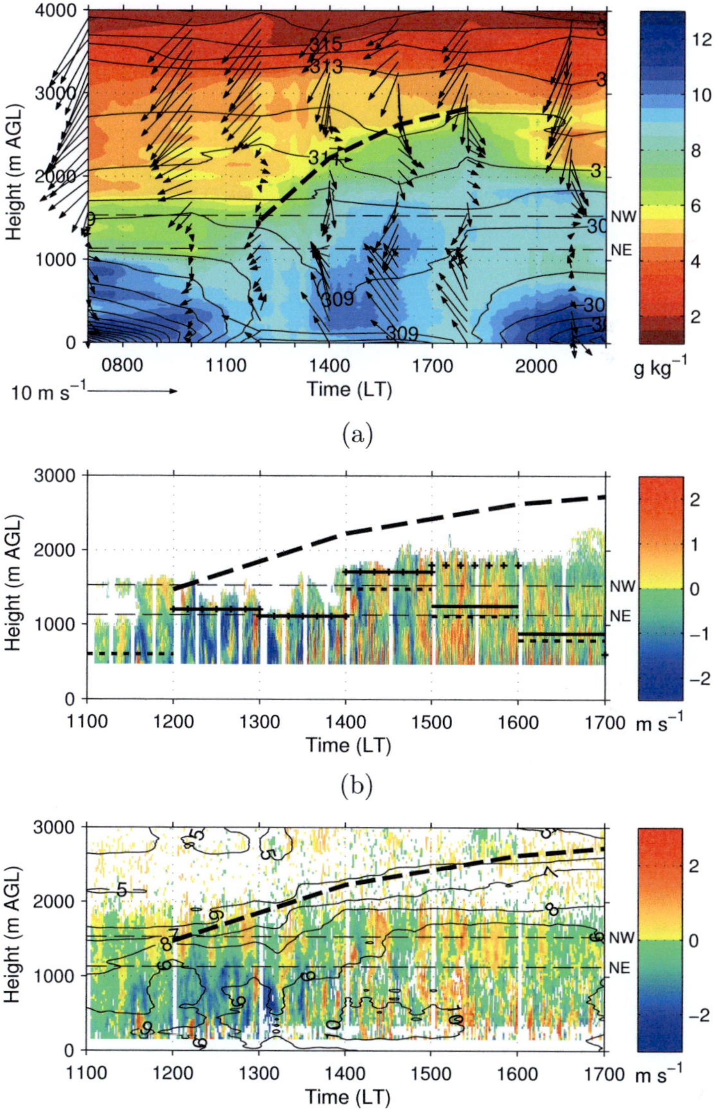

Height (m AGL)
Time (LT)
10 m s⁻¹
g kg⁻¹
NW
NE
315
313
311
309
0800
1100
1400
1700
2000

(a)

(b)

(c)

different beam characteristics, the cloud radar returned a signal in a deeper layer than the lidar and provided additional information. The elevated vertical motions consisted of alternating up- and downdraughts in a slightly stably stratified layer between approximately 1200 and 1800 m AGL, i.e. they occurred within parts of wind layer I and were not connected to surface-based convective cells. The vertical motions extended over the whole layer depth of about 600 m. The duration of the up- and downdraughts was often about 15 minutes and vertical wind speeds of up to 1 m s^{-1} were measured in the updraughts, while the wind speed in the downdraughts was slightly lower. Before the onset of the elevated vertical motion at around 1130 LT, the valid data supplied by the wind lidar and thus higher aerosol content were mainly restricted to below 1000 m AGL (Fig. 5.4b). Once the elevated vertical motions started, the aerosol content increased rapidly up to about 1500 m AGL and varied strongly with height afterwards. After

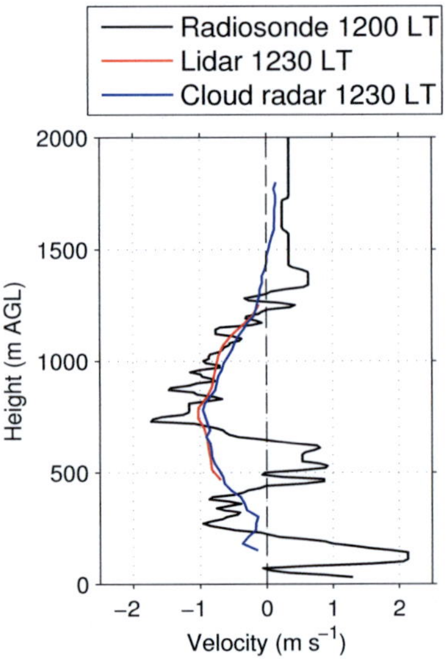

Figure 5.5.: Difference of the radiosonde ascent rate from its mean ascent rate at 1200 LT and mean vertical wind speed measured by a wind lidar and cloud radar between 1130 and 1230 LT at Corte on 19 August.

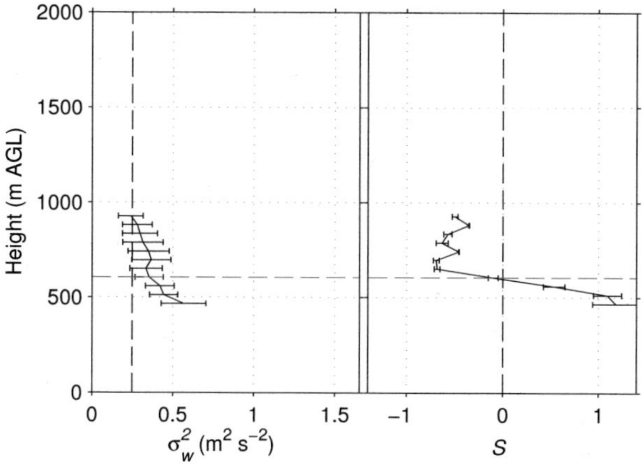

(a)

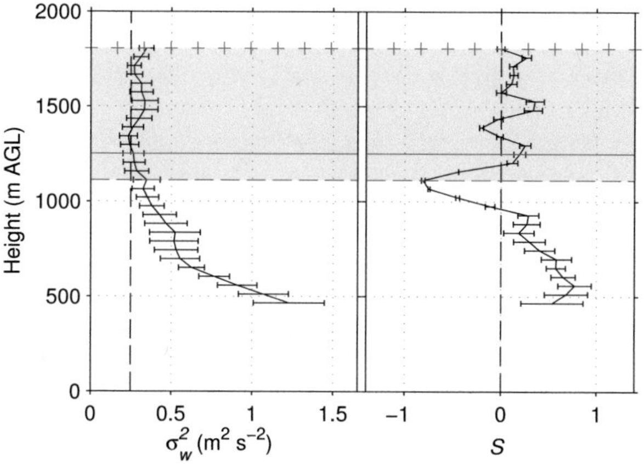

(b)

Figure 5.6.: Variance, σ_w^2, (left) and skewness, S, (right) of the vertical wind speed measured by a wind lidar at 1200 LT (a) and 1600 LT (b) at Corte on 19 August. The error bars show the total error; the vertical dashed black line indicates the used variance threshold and the zero skewness line, respectively; the horizontal grey solid line marks the variance- and skewness-based convection-layer depth; and the grey dashed line marks the lower and the grey crosses mark the upper boundaries of the uncertainty range (shaded), if present.

around 1300 LT, alternating up- and downdraughts still existed in wind layer I, but the vertical extent of the updraughts was no longer confined to this layer (Figs. 5.4b and c). Surface-based convective cells occasionally coupled to the elevated updraughts, which resulted in updraughts extending from near the surface up to about 1800 m AGL, for example between about 1310 and 1320 LT, 1400 and 1430 LT and 1445 and 1530 LT. Especially the first few of these coupling periods were accompanied by an increase of humidity (Fig. 5.4c) and aerosol content (Fig. 5.4b) in the layer with elevated vertical motions. During periods with coupling, the variance was rather high and the skewness was mostly positive throughout all layers with valid data (Fig. 5.6b). A weak local σ_w^2 maximum occurred between about 1300 and 1700 m AGL. At 1500 LT, a convection-layer height of about 1700 m a.g.l. was reached and at 1600 LT, it ranged between about 1100 and 1800 m a.g.l. (Figs. 5.4b and 5.6b). The simultaneous measurements of humidity and vertical wind speed revealed that the convection layer in the Tavignano Valley at Corte was often significantly lower than the mountain ABL (Fig. 5.4b).

Shortly after the surface sensible heat flux became positive (Fig. 5.3), the nocturnal surface inversion broke up and at 1200 LT, the potential temperature was constant with height up to about 1200 m AGL, like the specific humidity (Figs. 5.4a). This means that the height-constant profiles extended over a layer significantly deeper than the convection layer calculated from wind-lidar data between 1100 and 1200 LT (Fig. 5.4b). Until 1400 LT, the potential temperature increased mainly between about 400 and 1800 m AGL, i.e. within parts of the upvalley-wind and convection layer and within wind layer I, resulting in a stably stratified layer (Fig. 5.4a). Below that layer, a neutrally stratified layer of about 300 m depth and an unstably stratified layer of about 100 m depth adjacent to the surface persisted. After 1400 LT, a slight increase of potential temperature occurred mainly within wind layer II. After around 1900 LT, a nocturnal surface inversion developed again.

Based on these observations, the following questions arose. They are addressed in the subsequent sections.

1. What was the reason of the strong subsidence in the lower 1200 m AGL between about 1130 and 1330 LT?

2. What caused the temperature and humidity changes in wind layers I and II?

3. What was the reason of the elevated vertical motions?

4. Why did surface-based convective cells occasionally couple to elevated updraughts? How did this affect the conditions in the mountain ABL?

5. Which processes were mainly responsible for the evolution of the mountain ABL?

5.1.2. Processes in the Mountain ABL

Valley Wind and Sea Breeze

On the east coast as well as on the west coast of the Corsican Island, a sea breeze developed in the morning (Fig. 5.7). The arrival of the sea-breeze front was indicated by a wind-speed acceleration and a shift to south-westerly directions at Ajaccio on the west coast and to easterly directions at San Giuliano and Aléria on the east coast. Between 0900 and 1000 LT, a sea breeze started at both coasts and soon reached wind speeds of more than 4 m s^{-1} at Aléria and Ajaccio. At San Giuliano, it remained somewhat weaker. Although the surface sensible heat flux reached values higher than 400 W m^{-2}, as measured at San Giuliano, the near-surface temperature on the coasts increased only slightly after the sea-breeze onset, likely due to the horizontal advection of colder air from the sea. Simultaneously with the sea-breeze front arrival near the surface (Fig. 5.7), the wind turned to easterly in a layer of at least 300 m depth, as detected by sodar measurements at San Giuliano (Fig. 5.8a). Because of large-scale north-easterly wind direction it was difficult to determine the top of the stably stratified sea-breeze layer via the wind direction. Instead, a strong gradient of humidity and temperature at around 700 m AGL at 1200 LT was interpreted as the sea-breeze layer height (Fig. 5.8b). This was similar to the sea-breeze layer depth at Ajaccio at 1300 LT, identified as the layer with high humidity and south-westerly wind. A climatological study has revealed somewhat higher typical sea-breeze layer depths of around 1000 m at Ajaccio (Späth, 2012). While the sea breeze on the west coast was associated with moister air near the surface (Fig. 5.7), the specific humidity within the

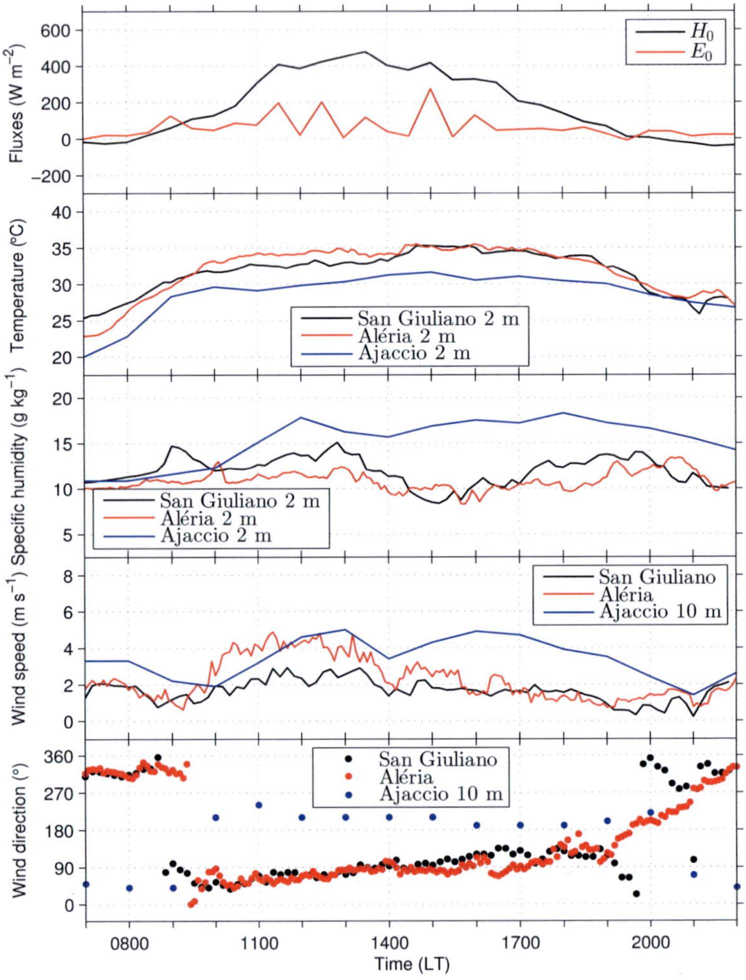

Figure 5.7.: Surface sensible heat flux, H_0, and latent heat flux, E_0, at San Giuliano; air temperature and specific humidity; horizontal wind speed; and wind direction in different heights at San Giuliano and Aléria on the east coast and at Ajaccio on the west coast on 19 August.

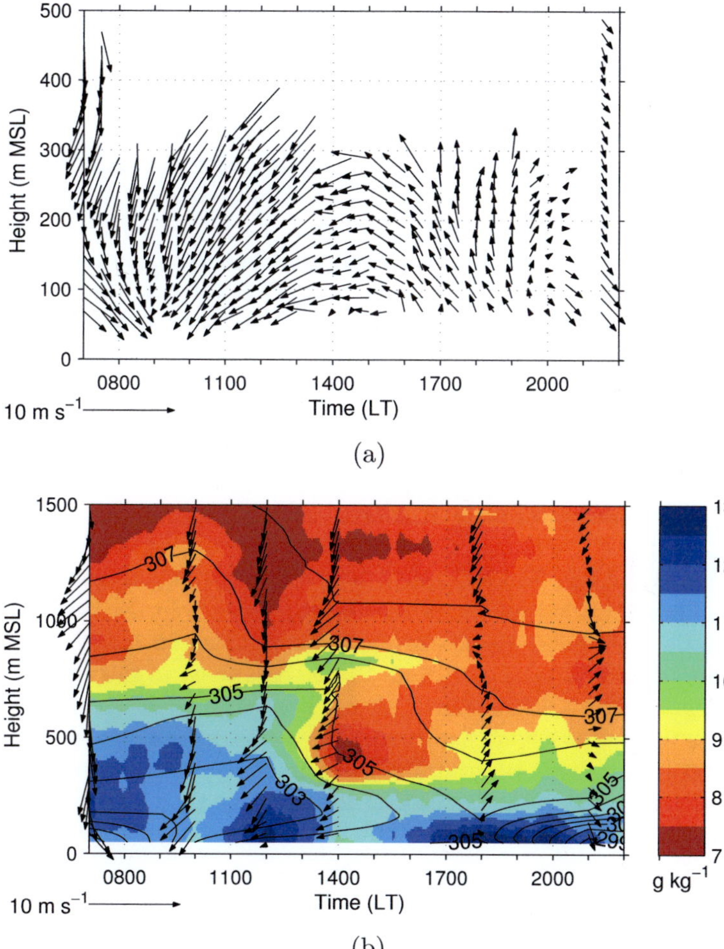

Figure 5.8.: Horizontal wind vector measured by sodar (a) and specific humidity (colour-coded), potential temperature (black isolines) and horizontal wind vector (arrows) measured by radiosondes (b) at San Giuliano on 19 August. Specific humidity is interpolated using microwave radiometer and radiosonde data and potential temperature is temporally linearly interpolated between the radiosoundings.

sea breeze on the east coast decreased in about the lower 700 m MSL after about 1300 LT (Figs. 5.7 and 5.8b). This was probably caused by the advection of drier air from the north-east due to large-scale humidity gradients. After around 1600 LT, the humidity increased again in the lowest 500 m. While the sea breeze was maintained below about 150 m MSL until about 2000 LT, the wind above already gained a westerly component after around 1800 LT, probably indicating the formation of a land breeze.

After the sea breeze had developed in the morning, it likely penetrated the interior of the island. The humidity increase in the upvalley-wind layer at around 1230 UTC at Corte (Figs. 5.3 and 5.4a) was attributed to the arrival of the sea-breeze front, which implies the formation of a combined upvalley-wind and sea-breeze layer in the Tavignano Valley. The arrival was connected with a wind-speed increase of more than 2 m s^{-1} close to the surface (Fig. 5.3). Along the valley axis, the distance between Corte and Aléria was about 35 km. Since the sea-breeze front arrived at Corte about 3 hours later than at Aléria, a mean propagation speed of about 3 m s^{-1} was estimated, which is slightly faster than values reported by e.g. Finkele (1998). The estimated propagation speed was less than half of the observed wind speeds in the combined upvalley-wind and sea-breeze layer (Figs. 5.3 and 5.4a). Once the drier air affected the conditions on the east coast in the afternoon (Figs. 5.7 and 5.8b), the drier air mass was also inland transported with the upvalley wind and reached Corte between about 1500 and 1600 UTC (Figs. 5.3 and 5.4a). From the existing data, it was not clear whether the increase in the upvalley wind-layer depth to 1000 m between 1200 and 1400 LT was exclusively caused by the sea breeze.

Slope Winds

The typical onset and duration of upslope winds near the surface were analysed for data measured at Rusio, a station on a north-east-facing slope. At around 0800 LT, the wind turned and gained an easterly upslope component (Fig. 5.9). Its strength varied and maximum values of almost 4 m s^{-1} were reached around noon. After around 1800 LT, a diffuse transition to downslope wind occurred.

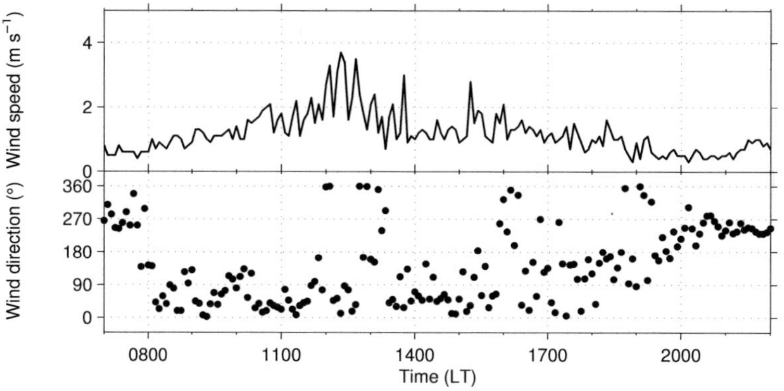

Figure 5.9.: Horizontal wind speed and wind direction measured near the surface at Rusio on 19 August.

At Corte, elevation scans perpendicular to the valley axis and azimuth scans were performed with the wind lidar to analyse the depth and extent of the upslope winds. After around 0930 LT, flow away from the lidar was detected near the slopes, reflecting upslope winds in a layer of about 100 m depth. Until 1200 LT, the depth of the upslope wind components at 225° and 45° increased up to about 500 m (Fig. 5.10a). Similar upslope-flow depths were reported by Reuten et al. (2005). For the azimuth scan, the mean radial velocity, $\overline{rv}$, was determined for each radius by fitting a cosine to the measured radial velocity, rv. By subtracting $\overline{rv}$, the remaining radial velocity $rv' = rv - \overline{rv}$ provided information on spatial inhomogeneities of the radial velocity field. At 1200 LT, rv' clearly showed two directions with positive values (Fig. 5.10b), which were associated with the slopes north-east and south-west of Corte and indicated the horizontal extent of the upslope winds. The radial velocities were enhanced near the juncture between a tributary valley and the main valley south-west of Corte (Fig. 5.10b). This might be related to an upvalley wind forming in the tributary valley.

Subsidence in the Valley

The upslope winds developing in the Tavignano Valley presumably caused divergence in the lower part of the valley atmosphere

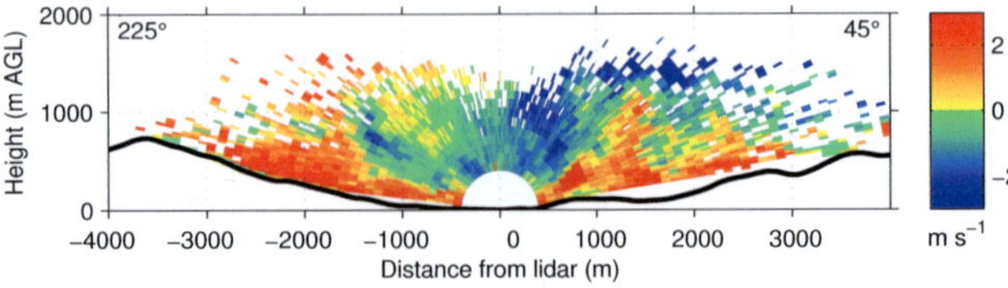

(a)

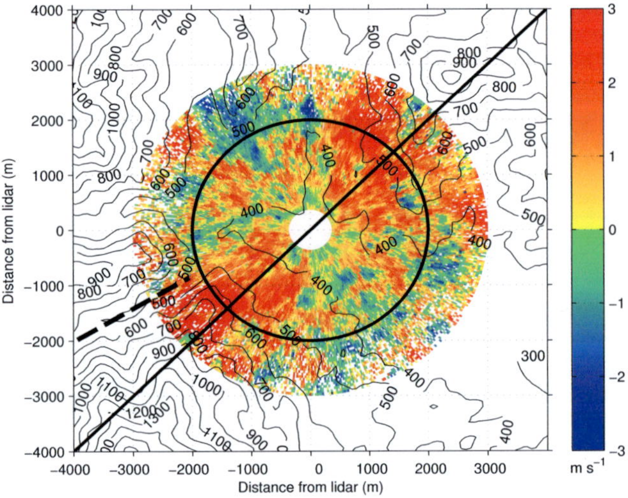

(b)

Figure 5.10.: Radial velocity during an elevation scan perpendicular to the valley axis (a) and radial velocity minus mean radial velocity ($rv' = rv - \overline{rv}$) during an azimuth scan at 10° elevation angle (b) measured by a wind lidar at Corte at around 1200 LT on 19 August. Positive radial velocity indicates flow away from the lidar and negative radial velocity indicates flow towards the lidar. In (a), azimuth directions are indicated in the upper left and right corners and the orography along the scan is shown by the black line. In (b), isolines denote the orography in m MSL, the black slolid line shows the direction of the elevation scan and the black dashed line marks the axis of a tributary valley south-west of Corte. The black circle has a radius of 2000 m.

and hence might have been responsible for the strong subsidence measured in the valley centre between about 1130 and 1330 LT (Figs. 5.4b, 5.4c and 5.10a). To verify this assumption, the strength of the divergence was estimated using the wind-lidar azimuth scans at 10° elevation angle (Fig. 5.10b). Assuming a divergence-free valley atmosphere, the mass balance was calculated for a cylinder volume centred at Corte with volume, V_c, surface, S_c, radius, R_c, and height, H_c. For the following calculations, the radial velocity was projected onto the x-y plain: $rv_{xy}(r_c, \phi) = rv(r_c, \phi)\ (\cos(10°))^{-1}$, where r_c is the distance from the cylinder centre and ϕ is the azimuth angle with 0° pointing north. In the following calculations, rv_{xy} is assumed to be height-constant within the cylinder. The Gauss' theorem

$$\int_{V_c} \nabla \cdot \vec{v}\ dV_c = \oint_{S_c} \vec{v} \cdot \vec{n}\ dS_c = 0 \tag{5.1}$$

in cylinder coordinates was used to describe the mass flux through the cylinder surface with the unit normal vector

$$\vec{n} = (\vec{e}_{r_c}, \vec{e}_{\phi}, \vec{e}_z), \tag{5.2}$$

where $\vec{e}_{r_c}$, $\vec{e}_{\phi}$, $\vec{e}_z$ are basis vectors, and the velocity vector

$$\vec{v} = \begin{pmatrix} rv_{xy}(r_c, \phi)\ \sin\phi \\ rv_{xy}(r_c,\ \phi)\ \cos\phi \\ w(r_c, h_c) \end{pmatrix} \tag{5.3}$$

where h_c is the height from cylinder bottom. It is further assumed that the flux at the cylinder bottom vanishes. The flux through the cylinder barrel (index B) with the radius R_c amounts to

$$\begin{aligned} F_{c,B} &= \int_{S_{c,B}} \vec{v} \cdot \vec{e}_{r_c}\ dS_{c,B} \\ &= \int_{S_{c,B}} \begin{pmatrix} rv_{xy}(R_c, \phi)\ \sin\phi \\ rv_{xy}(R_c, \phi)\ \cos\phi \\ w(R_c, h_c) \end{pmatrix} \cdot \begin{pmatrix} \sin\phi \\ \cos\phi \\ 0 \end{pmatrix} dS_{c,B} \\ &= \int_0^{2\pi} \int_0^{H_c} rv_{xy}(R_c, \phi)\ R_c\ dz d\phi \\ &= \frac{2\pi\ R_c\ H_c}{n} \sum_{i=1}^{n} rv_{xy}(R_c, \phi_i) \end{aligned} \tag{5.4}$$

with the number of angles during an azimuth scan, n. The flux through the cylinder top (index T) at height H_c is calculated to

$$\begin{aligned} F_{c,T} &= \int_{S_{c,T}} \vec{v} \cdot \vec{e}_z \, dS_{c,T} \\ &= \int_{S_{c,T}} \begin{pmatrix} rv_{xy}(r_c, \phi) \sin\phi \\ rv_{xy}(r_c, \phi) \cos\phi \\ w(r_c, H_c) \end{pmatrix} \cdot \begin{pmatrix} 0 \\ 0 \\ 1 \end{pmatrix} dS_{c,T} \qquad (5.5) \\ &= \int_0^{2\pi} \int_0^{R_c} w(r_c, H_c) \, r_c \, dr_c d\phi. \end{aligned}$$

Assuming a sine-shaped vertical wind speed distribution across the cylinder top

$$\begin{aligned} w(r_c, H_c) &= w_0 \cos\left(\frac{\pi}{2\, R_c} \, r_c\right) \\ w(0, H_c) &= w_0 \qquad (5.6) \\ w(\pm R_c, H_c) &= 0 \end{aligned}$$

the flux through the cyclinder top (Eq. 5.5) results to

$$F_{c,T} = 1.5 \, w_0 \, R_c. \qquad (5.7)$$

To conserve mass $F_{c,B} = -F_{c,T}$ has to apply and the strongest downward motion, w_0, through the cylinder top caused by the observed horizontal divergence at a certain radius amounts to

$$w_0 = -\sum_{i=1}^{n} rv_{xy}(R_c, \phi_i) \, \frac{d \, H_c}{n \, R_c}, \qquad (5.8)$$

with the scalar $d \approx 4.3$. The layer of maximum subsidence, measured by wind lidar in the vertical stare mode (Fig. 5.4b) and averaged for the two-hour period, was taken as cylinder height $H_c \approx 630$ m AGL, assuming that horizontal divergence existed below and horizontal convergence above this layer. This height roughly agreed with the observed depth of the upslope-wind layer (Fig. 5.10a). Next, w_0 was calculated for radii R_c ranging from 400 to 3000 m. For cylinders with radii larger than 2000 m (Fig. 5.10b), the strongest downward motion calculated was about $w_0 = -0.7$ m s^{-1}, which was in the same order of magnitude as the measured vertical wind speed. This means that an area with subsidence covering at least 4000 m of the valley width was possible to exist. This extent approximately coincided with the one measured during the wind-lidar elevation scan (Fig. 5.10a). These results indicate that the observed strong mean subsidence was in fact a part of the thermally driven circulations and not caused by other processes, e.g. by the impact of the background flow.

Advective Venting

Advective venting of humidity requires the evolution of horizontal humidity gradients due to the orography. In order to detect humidity inhomogeneities, data from a scanning microwave radiometer at Corte were used. Figure 5.11a shows the spatio-temporal evolution of IWV measured by the radiometer during azimuth scans. At around 0930 LT, first spatial differences in the IWV distribution in the order of less than 1 kg m^{-2} occurred. They quickly increased to about 3 kg m^{-2} at 1200 LT (Fig. 5.11). Three main directions with higher IWV values could be distinguished at around 1200 LT: 20° to 125°, 195° to 225° and 290° to 330°. The polar plot of IWV with the underlying orography at 1200 LT reveals that higher IWV values were mostly correlated with higher orography (Fig. 5.11b). In the directions of higher IWV, the distance between beam and underlying orography was mainly below 500 m within a range of 2000 m from Corte. Combined with wind-lidar measurements (Fig. 5.10), this suggests that humidity in the upslope-wind layer contributed to the higher IWV probably together with convection over the slopes and ridges. In the afternoon, the difference between IWV measured by the microwave radiometer in the valley during the vertical stare mode (Fig. 5.3) and the IWV above the higher terrain reached more than 3 kg m^{-2} (Fig. 5.11a). Once the humidity in the upvalley-wind layer at Corte increased at around 1230 LT (Fig. 5.4a), probably associated with the arrival of the sea-breeze front, the absolute IWV values increased in all directions (Fig. 5.11a). At that time, first clouds developed above the mountain ridges surrounding Corte and reached their maximum intensity at around 1500 LT but remained shallow, as was seen by a cloud camera (Fig. 5.12). The clouds existed until around 1630 LT. Their decay correlated with a decrease of the humidity above the

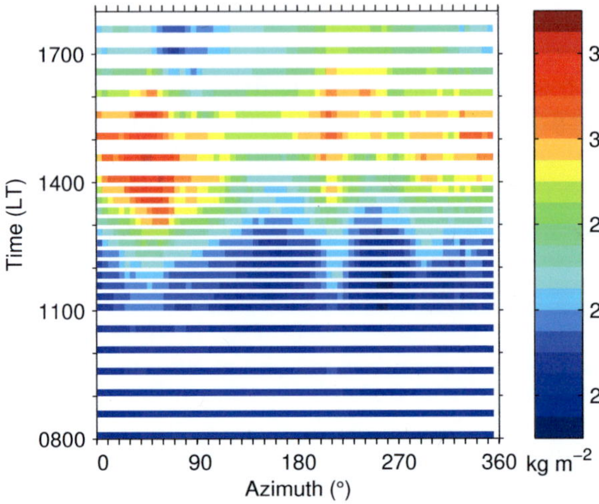

(a)

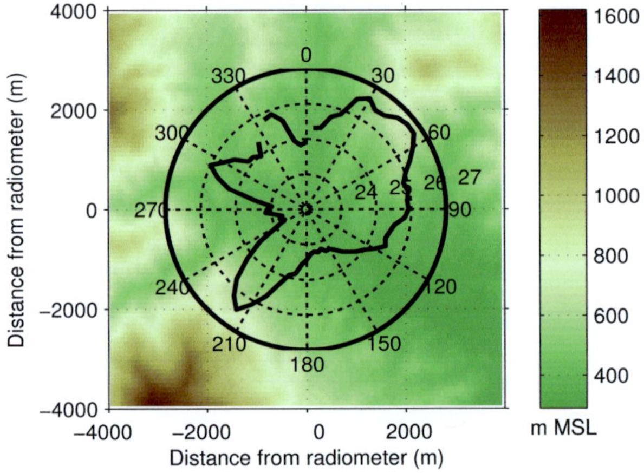

(b)

Figure 5.11.: Azimuth time plot of IWV (a) and polar plot of IWV with underlying orography at 1200 LT (b) at Corte on 19 August for an elevation angle of 19.8°. IWV was measured by a microwave radiometer.

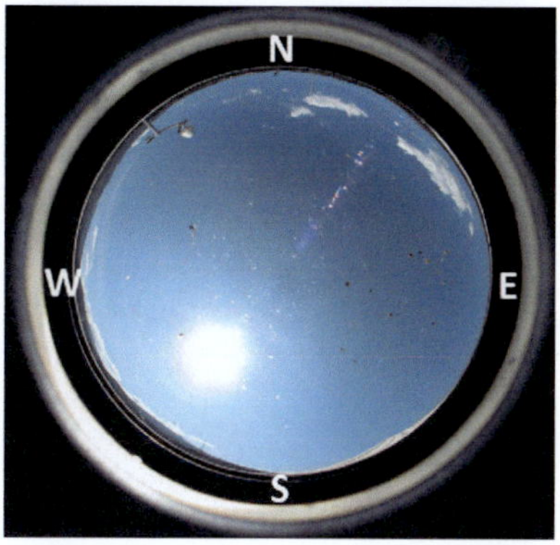

Figure 5.12.: Fisheye image taken by a cloud camera at Corte at around 1500 LT on 19 August.

slopes and ridges (Fig. 5.11a) and a weakening of upslope winds measured by a wind lidar.

Further information on horizontal inhomogeneities above ridge height were obtained by aircraft measurements. As no aircraft measurements were available on 19 August, data from 23 September were analysed, another day with an undisturbed mountain ABL evolution (Fig. B.4 in the appendix). During the afternoon, the aircraft perpendicularly crossed the mountains west of Corte, with the distance between aircraft and ridges partly being less than 500 m. Above a ridge at around 8.86° E, positive vertical wind speed of several m s^{-1}, higher specific humidity, lower potential temperature and lower horizontal wind speed than in the surroundings were measured (Fig. 5.13). This suggests the existence of a layer influenced by convection and horizontal convergence over the ridge. While the evolution of a CBL above a ridge typically leads to a higher humidity content compared to the ambient air at the same height above the valley, the temperature above the ridge may be either higher or lower than above the valley, depending on the vertical distance from the mountain ridge; e.g. the upper part of the layer is cooler than the ambient air. Assuming that the same processes occurred on 19 August, horizontal humidity and temperature gradients were likely to develop due to convection and topographic venting and to allow the occurrence of advective venting.

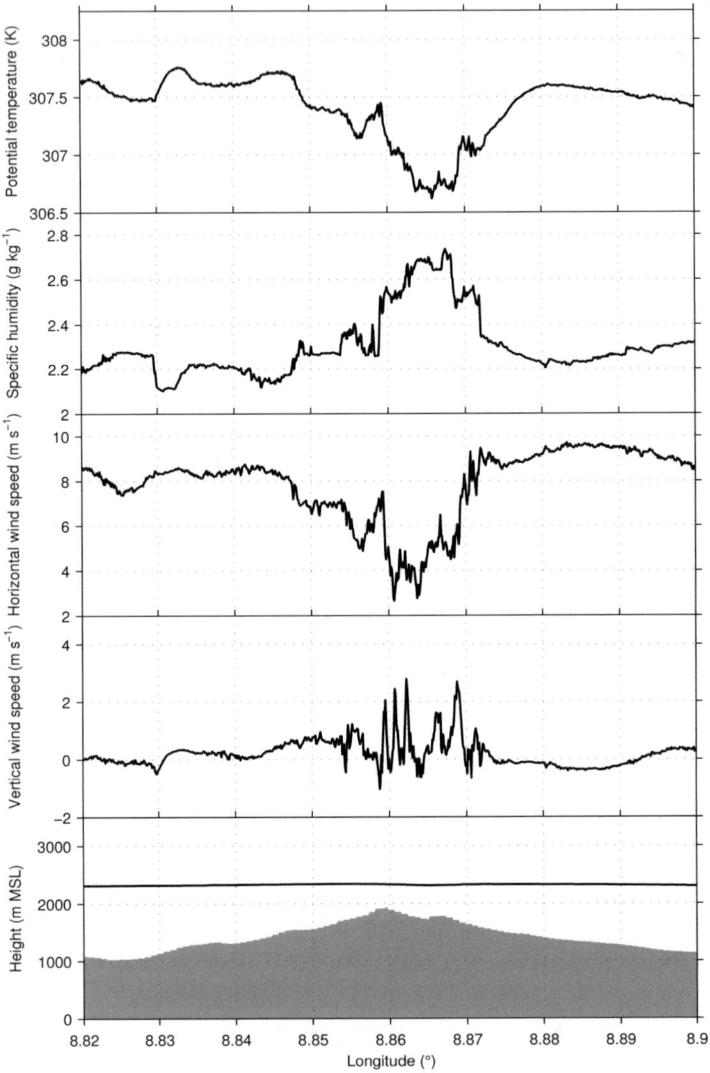

Figure 5.13.: Potential temperature, specific humidity, horizontal wind speed and vertical wind speed as well as flight level and orography along a zonal flight track at about 42.3° N at around 1400 LT on 23 September.

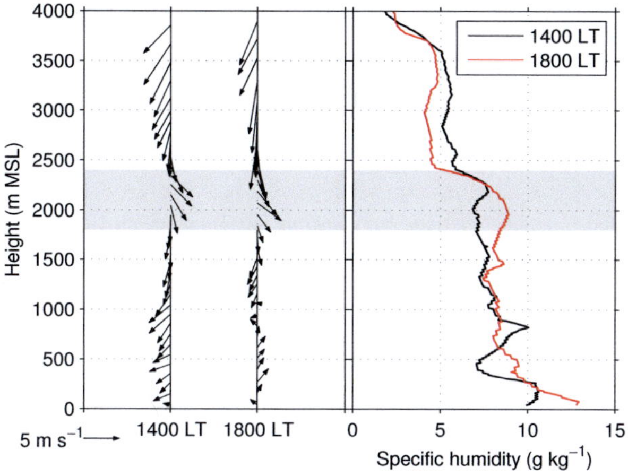

(a)

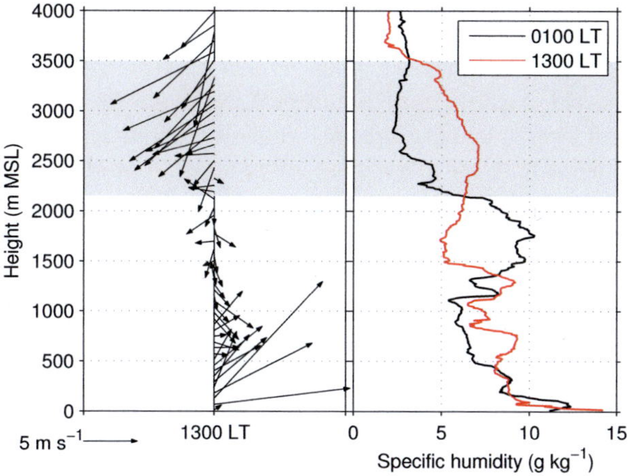

(b)

Figure 5.14.: Horizontal wind vectors and specific humidity measured by radiosondes at San Giuliano (a) and Ajaccio (b) on 19 August. Grey shading marks the layers presumably affected by advective venting.

The horizontal flow in wind layer II (Fig. 5.2) allowed for the transport of water vapour to the valley centre from the mountain ridges north-west of Corte, where higher humidity existed, as was evident from radiometer measurements (Fig. 5.11). Below this layer, wind layer I with northerly to easterly winds (Fig. 5.2) existed during the day and also allowed for the transport of higher humidity from the mountains north-east of Corte to the valley centre (Fig. 5.11). Thus, advective venting obviously contributed the humidity increase within wind layers I and II in the valley (Fig. 5.4a).

Humidity changes in elevated layers also occurred on the east and west coasts of the Corsican Island, several kilometres away from the mountains. At San Giuliano on the east coast, a layer with wind directions between 210° and 330°, i.e. wind blowing from the mountains towards the coast, developed between about 1800 and 2400 m MSL in the afternoon (indicated by the grey shading in Fig. 5.14a). Within parts of this layer, the humidity increased by up to 2 g kg^{-1} between 1400 and 1800 LT. At Ajaccio, a westerly wind component in the lowest 1500 m MSL (Fig. 5.14b) indicated the evolution of a thermally driven circulation on the steep west coast, which probably was a combination of sea breeze, upslope wind and upvalley wind. The specific humidity increased by up to 4 g kg^{-1} in a layer between about 2200 and 3500 m MSL at 1300 LT compared to the previous night (indicated by the grey shading in Fig. 5.14b). The increase occurred in a layer where north-easterly wind prevailed, i.e. where air was advected from above the mountains towards the coast. The relation between the wind directions and the humidity increases on the east and west coasts suggests that these increases were also caused by advective venting.

Elevated Vertical Motion

The elevated vertical motions, which occurred mainly in wind layer I (Figs. 5.2, 5.4b and 5.4c), might be explained by two different processes: (i) convection and non-stationary upslope winds (Fig. 5.9) might have caused intermittent updraughts above the ridges, which might have been horizontally advected with the mean north-easterly wind. (ii) Internal gravity waves initiated by orography might have occurred in the stably stratified layer. The first explanation seems more likely, because the elevated vertical motions were only observed during the period of upslope winds and their frequency was not affected by any changes in stability during the few hours of interest. This hypothesis is also supported by the rather high turbulence (Fig. 5.6b) and the increase of aerosol content in the relevant layer (Fig. 5.4b), as a higher aerosol content usually suggests that the air was in contact with the Earth's surface. This is in agreement with previous studies, in which the advection of elevated aerosol layers (Kossmann et al., 1999; Gohm et al., 2009) or elevated mixed layers (Arritt et al., 1992; Sun and De Wekker, 2011) was observed and advection of elevated turbulence from nearby slopes in the valley centre was simulated (Serafin and Zardi, 2010a,b).

As long as the convection layer in the valley centre did not reach the layer with elevated vertical motions, no connection between elevated updraughts and surface-based convective cells was found (Figs. 5.4b and c). Only after the convection layer had grown up to the bottom of the layer with elevated vertical motions did surface-based convective cells occasionally couple to the elevated updraughts, which resulted in a strong growth of the convection-layer depth. The high σ_w^2 (Fig. 5.6b) and the rapid increase of humidity (Fig. 5.4c) and aerosol content (Fig. 5.4b) in the layer with elevated vertical motions, which were associated with the coupling, suggest that deep vertical turbulent mixing occurred and humidity and aerosols were transported upwards. This indicates that cell coupling was an effective vertical transport mechanism which resulted from the superposition of processes on different scales, i.e. turbulent convective cells and mesoscale advection with the mean wind.

Diurnal Cycle of IWV

The impact of humidity changes in different layers in the mountain ABL on the IWV evolution was already noted at Corte in Sect. 5.1.1. Here, this relation is analysed in more detail for Corte, San Giuliano and Ajaccio. Evapotranspiration at the Earth's surface contributes to an increase of IWV (Sect. 2.5). A surface latent heat flux of 100 W m^{-2} lasting for 6 hours, which approximately coincided with the measurements at San Giuliano (Fig. 5.7), would cause an IWV increase of about 1 kg m^{-2}. Thus, it is assumed that

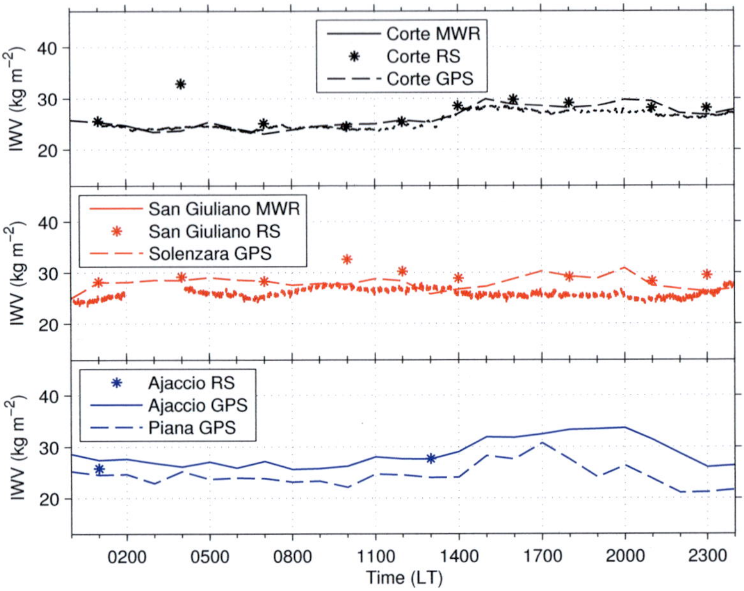

Figure 5.15.: IWV measured by microwave radiometers (MWR), radiosondes (RS) and GPS at Corte, the east coast and the west coast on 19 August.

rises of IWV larger than this are related to mesoscale processes and to the evolution of a deep mountain ABL, in the absence of large-scale processes or phase changes.

At Corte, the IWV increase by more than 5 kg m^{-2} after around 1230 LT (Fig. 5.15) was attributed to the rise of humidity in the upvalley-wind layer and in wind layers I and II (Fig. 5.4a). The maximum IWV values at Corte were reached at around 1600 LT, although the moistening in wind layer II lasted till about 1800 LT. Its contribution to the high IWV was possibly masked by the advection of dry air with the upvalley wind from the east coast in the second half of the afternoon.

The slight increase of IWV at San Giuliano until around 0930 LT (Fig. 5.15) was likely related to the transport of moist air with the sea breeze on the east coast (Figs. 5.7 and 5.8b). The humidity increase in the elevated layer due to advective venting at San Giuliano in the afternoon (Fig. 5.14a) was not reflected in the IWV measurements (Fig. 5.15), probably because it was obscured by large-scale advection of dry air in the sea-breeze layer (Fig. 5.8b) and in a layer between about 2300 and 3600 m MSL during the afternoon (Fig. 5.14a). A similar behaviour of the temporal evolution of IWV was found further south on the east coast, at Solenzara (Fig. 5.15).

The IWV at Piana and Ajaccio on the west coast showed strong diurnal cycles of about 8 kg m^{-2} with maximum values in the late afternoon or early evening (Fig. 5.15). As the near-surface wind speed and humidity in the sea breeze were rather constant throughout the day (Fig. 5.7), it was assumed that no significant changes of humidity, strength or depth in the sea-breeze layer occurred that could explain the rising IWV in the early afternoon. It was concluded that advective venting of water vapour from the mountains in the layer with north-easterly wind (Fig. 5.14b) dominated the IWV evolution and led to the strong increase in the afternoon.

The observations suggest that the evolution of a deep mountain ABL due to topographic and advective venting was associated with a distinct diurnal cycle of IWV: it increased in the early afternoon, reached maximum values in the the late afternoon or early evening and decreased afterwards.

5.1.3. Conceptual Model

At Corte in the Tavignano Valley, convection and mesoscale advection co-existed and affected the evolution and structure of the mountain ABL. To illustrate temporal and spatial interactions of the relevant processes, a conceptual model was developed (Fig. 5.16). The schematic diagrams summarize the processes for 19 August; the relative importance of the different processes may vary on other days with undisturbed mountain ABL evolutions.

In the morning, upslope winds and somewhat later, an upvalley wind in the Tavignano Valley developed (Fig. 5.16a). Shortly after noon, the humidity in the upvalley-wind layer increased, which was attributed to the arrival of the sea-breeze front at Corte.

In late morning, strong subsidence in the order of -1 m s^{-1} lasting for several hours was detected in the valley centre below the ridge heights (Fig. 5.16a). The estimation of the mass balance for the

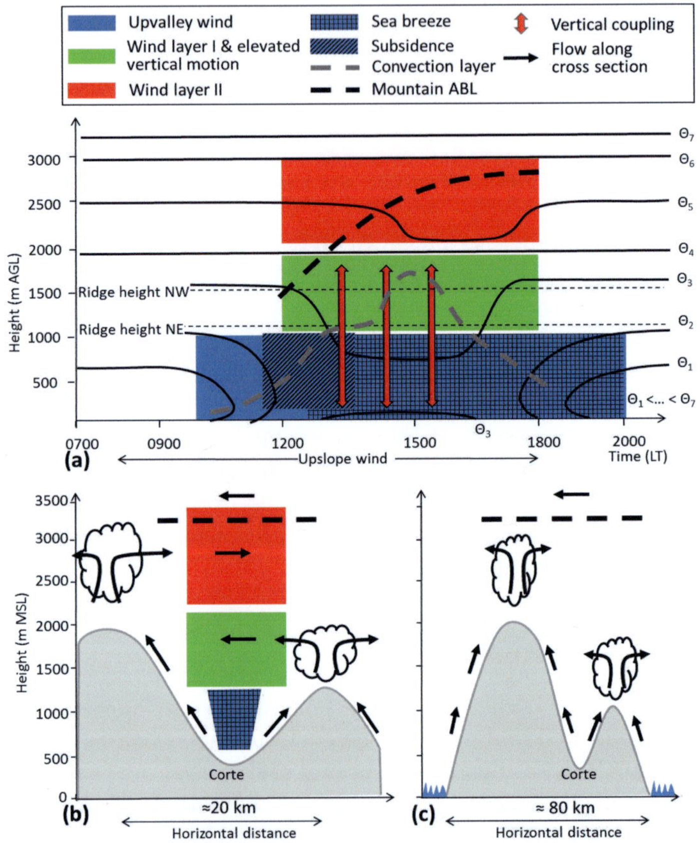

Figure 5.16.: Conceptual model of the temporal evolution of the conditions at Corte (a) and of the spatial processes along a west-east cross section in the afternoon (b) and (c) on different scales.

valley revealed that this mass flux compensated low-level divergence due to thermally driven flows. In most previous observational studies mean subsidence in a valley atmosphere caused by thermally driven circulations was not measured directly, but estimated from other observations. Mean subsidence on the scale of centimetres per second was obtained (Whiteman, 1982; Kondo et al., 1989). Aircraft measurements in a steep and narrow Alpine valley close to a slope revealed a similarly strong subsidence (Weigel and Rotach, 2004; Rotach et al., 2008). However, these authors assumed that this was enhanced by a curvature-induced secondary circulation (Sect. 2.3). Numerical simulations revealed subsidence in the range of -0.1 to -0.3 m s^{-1} (e.g. Rampanelli et al., 2004; Bischoff-Gauß et al., 2008; De Wekker, 2008; Serafin and Zardi, 2010a). At Corte, the measured strength of the mean subsidence was unexpectedly strong and had, to the author's knowledge, not been observed by continuous remote sensing at a valley site before.

Convective cells from the valley floor extended into the layer with subsidence, and a convection layer of several hundred metres of depth developed (Fig. 5.16a). Neglecting horizontal advection, vertical motion and diabatic processes and assuming a constant capping inversion strength, a one-dimensional model described in Kalthoff et al. (2011) was applied to compare the convection-layer growth in the valley to the growth of a CBL over homogeneous terrain under equal surface forcing. Under the present conditions, the model calculated a CBL depth of more than 1500 m until 1200 LT, i.e. the CBL over homogeneous terrain at this time would be about 900 m deeper than the observed convection layer but about as deep as the detected mountain ABL in the valley. This indicates that the mean compensating subsidence strongly reduced the growth of the convection layer in the morning. On the other hand, the mean subsidence warmed the stably stratified air in the valley centre from above and presumably contributed to the breakup of the nocturnal surface inversion in the valley. This double impact of subsidence on the breakup of a nocturnal surface inversion was also described by e.g. Whiteman (1982) and Kondo et al. (1989). The layer in the mountain ABL at 1200 LT in which the potential temperature and specific humidity were constant with height up to about 1200 m AGL might have been a result of combined convection, mean subsidence and horizontal advection. It is assumed that a warm layer formed close to the slopes and above the ridges due to convection, which allowed for horizontal advection of heat with a wind component parallel to the temperature gradients. Within the layer with height-constant potential temperature, deeper convective cells developed and the convection layer increased by more than 500 m between 1200 and 1300 LT (Fig. 5.16a).

From late morning until early evening, vertical motions occurred in an elevated layer of about 600 m depth, which was roughly confined to wind layer I (Fig. 5.16a). As long as the convection layer did not extend to the bottom of the layer with elevated vertical motions, surface-based convective cells were not connected to the elevated updraughts. Once the vertical distance between the two layers was overcome at 1300 LT due to the growth of the convection layer, surface-based convective cells occasionally coupled to the elevated updraughts. Cells reaching from the surface to about 1800 m AGL were associated with another rapid increase of the convection-layer depth by more than 500 m between 1400 and 1500 LT. The cells possibly transported humidity and aerosols upwards from near the surface, which contributed to the observed increase in humidity and aerosol content in the layer with elevated vertical motions.

The observations suggest that the convection-layer evolution in the valley centre was strongly influenced by mesoscale processes: subsidence reduced its growth, while a previous labilisation due to advection or vertical cell coupling caused a rapid increase. Despite vertical mixing, warming due to mean subsidence and horizontal advection probably caused a stabilisation in large parts of the convection layer shortly after noon (Fig. 5.16a). The stable stratification persisted in the afternoon although the mean vertical wind speed was close to zero. Possible reasons for the reduced mean subsidence in the valley atmosphere were a compensation by convective activity or a weakening of the upslope winds. Within the combined upvalley-wind and sea-breeze layer, a stable stratification might have been maintained by horizontal advection, as the sea-breeze layer on the coast was stably stratified.

Horizontal differences of humidity developed, with higher humidity values and clouds above the ridges (Figs. 5.16b and c). The humidity gradients were probably caused by convection over the ridges and topographic venting. Serafin and Zardi (2010a) found that the top of the atmospheric layer that is disturbed over the ridges corresponds well to the level of neutral buoyancy of parcels rising from the mountain ridges. These authors suggested that the level of neutral buoyancy depends on mountain-ridge height, near-surface temperature and atmospheric stratification. At Rusio, the maximum near-surface potential temperature reached about 312 K. Based on the atmospheric stratification measured by radiosondes at Corte, this resulted in a level of neutral buoyancy of roughly 3200 m MSL, i.e. about 2800 m above Corte. This indicates that deep convective layers with higher humidity compared to the surroundings developed above the mountain ridges.

It is assumed that the complex flow structure at Corte and the evolution of the layer with a westerly wind component on the east coast were parts of an orographically induced circulation system. Wind layers I and II in the Tavignano Valley at Corte (Figs. 5.16a and b) possibly represented return flows of thermally driven circulations, as the onset and duration of the two layers were related to the evolution of the upvalley wind, upslope wind and sea breeze, but delayed. As the mountains north-west of Corte were higher than those in the north-east, this might explain why wind layer II was confined to a higher elevation. Depending on the wind direction in the respective elevated layer, advective venting of moist air from the ridges might occur and explain the elevated moist layers, which formed above the valley at Corte. Thus, it is concluded that advective venting in wind layers I and II affected the humidity distribution in the valley centre and contributed to the evolution of a mountain ABL at Corte up to 2800 m AGL (about 3200 m MSL), i.e. the mountain ABL depth considerably exceeded the convection-layer depth in the Tavignano Valley (Fig. 5.16a) and well agreed with the above estimated level of neutral buoyancy. The latter indicates that the mountain ABL height depended on mountain-ridge height, near-surface temperature and atmospheric stratification. By comparison, the CBL depth over homogeneous terrain calculated with the one-dimensional model would amount to about 2200 m in the afternoon. This means that the mountain ABL in the afternoon was deeper than the layer that would evolve due to convection over homogeneous terrain under equal surface forcing.

Like at Corte, the return flow of the probably combined upslope wind and sea breeze on the east coast might modify the large-scale flow and might cause the elevated moist layer due to advective venting. On the west coast, advective venting might occur with the large-scale flow. A return flow was not detected because of the similar large-scale wind direction. With respect to the definition of a mountain ABL, deep mountain ABLs also formed on the coasts of the Corsican Island when an elevated temporal humidity increase occurred, which was related to the island orography.

5.2. The Interrupted Mountain ABL Evolution: Case Study of 02 October 2012

5.2.1. Large-Scale Conditions

On 02 October, a mid-level trough over western Europe extended south over northern France and caused persistent westerly flow throughout the troposphere in the investigation area. Upstream of the Corsican Island, inversions at about 1500 and 2500 m MSL confined an elevated layer, hereafter named flowing layer, around the characteristic mountain-ridge height with a mean westerly wind of $U \approx 8\ \mathrm{m\ s^{-1}}$ and a mean potential temperature of $\Theta_\mathrm{m} \approx 299$ K at 1300 LT (Fig. 5.17). The upstream air mass below 1500 m MSL was blocked by the mountain ridge as indicated by a non-dimensional mountain height (Eq. 2.7) of $\hat{H} = 0.009\ \mathrm{s^{-1}} \cdot 1250\ \mathrm{m} \cdot (4\ \mathrm{m\ s^{-1}})^{-1} = 2.8$, which was calculated for mean layer values. For the flowing layer, the upstream Froude number of hydraulic flow theory (Eq. 2.25) was calculated to be $Fr \approx 8\ \mathrm{m\ s^{-1}} \cdot (10\ \mathrm{m\ s^{-1}})^{-1} = 0.8$.

In the investigation area, large-scale horizontal advection of dry and warm air occurred below about 4000 m MSL extending to lower levels in the course of the day. This decreased the IWV by at least 5 kg $\mathrm{m^{-2}}$ at Corte, San Giuliano and Ajaccio until the evening (Fig. 5.18) and resulted in subsiding and strengthening of the elevated temperature inversion at around 2500 m MSL (Fig. 5.17) during the afternoon.

To analyse mid-level conditions above the mountain ridge, aircraft measurements from two flight tracks perpendicular to the mountain ridge at about 3500 m MSL were available (Fig. 5.19), i.e. above the elevated temperature inversion. The temperature and vertical wind-speed waves showed a phase shift of about 90°, which revealed the existence of gravity waves (Sect. 2.4.2). As-

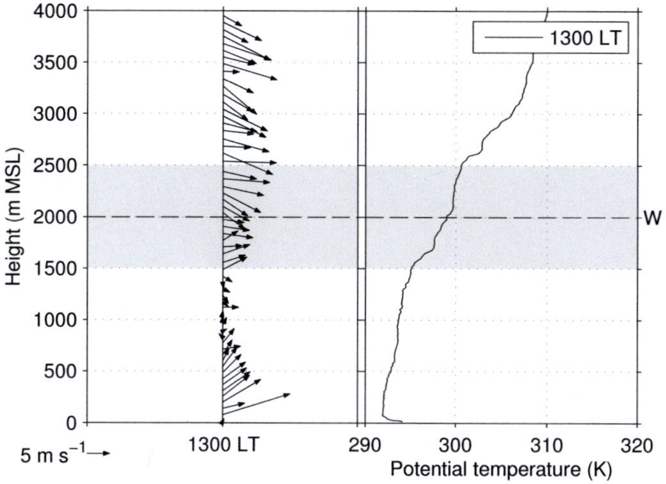

Figure 5.17.: Horizontal wind vectors and potential tempreature measured by radiosonde at Ajaccio at 1300 LT on 02 October. Grey shading marks the flowing layer and the characteristic mountain-ridge height to the west of Corte (W) is indicated by the black dashed line.

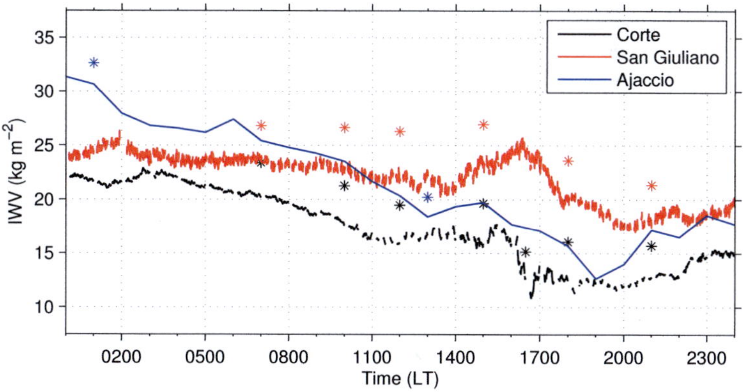

Figure 5.18.: IWV at Corte, San Giuliano and Ajaccio on 02 October. The stars indicate IWV calculated from radiosoundings and the solid lines mark IWV measured by radiometers at Corte and San Giuliano and GPS at Ajaccio.

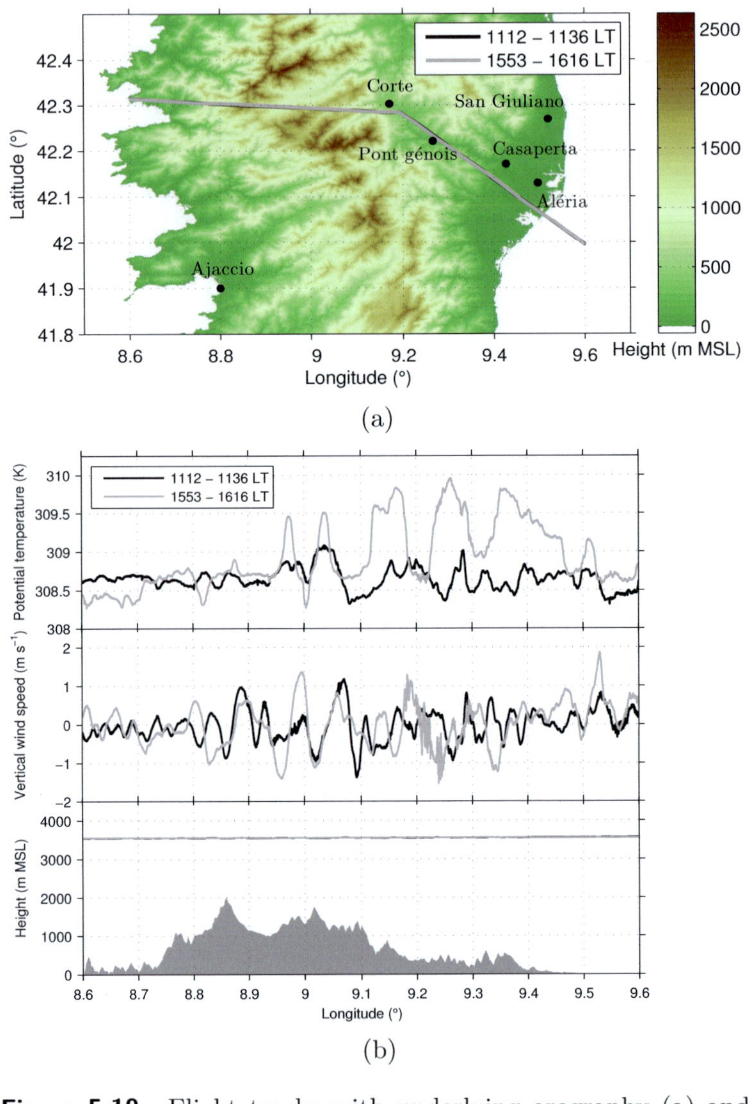

Figure 5.19.: Flight tracks with underlying orography (a) and corresponding potential temperature, vertical wind speed as well as flight level and orography along the flight tracks (b) on 02 October.

suming stationary conditions, wavelengths of mainly less than 10 km resulted. The amplitude of the vertical wind-speed wave was around 1 m s^{-1}. Along the flight tracks, the mountains are characterized by two successive ridges, at about 8.85° E and 9.02° E, separated by roughly 20 km and a valley in between, resulting in a terrain wavelength of about 20 km.

5.2.2. Mountain ABL Evolution on the Coasts

Near-surface measurements at Ajaccio on the west coast and at San Giuliano on the east coast revealed the evolution of a sea breeze during the day. Maximum sea breeze wind speeds reached 6 m s^{-1} on the west coast and 4 m s^{-1} on the east coast. Sodar measurements at San Giuliano captured the onset of the sea breeze at around 1230 LT up to about 400 m MSL (Fig. 5.20a). The easterly wind weakened after around 1700 LT, and turned to westerly wind at around 1930 LT. The maximum sea-breeze layer depth did not exceed 500 m MSL, as detected from wind profiles measured by radiosondes (Fig. 5.20b). The sea breeze top was accompanied by a strong humidity reduction.

In the course of the day, the humidity increased between the top of the sea-breeze layer and about 2300 m MSL by up to 2 g kg^{-1} in the afternoon (Fig. 5.20b). As this increase was mostly associated with westerly and partly southerly wind, a transport of water vapour from above the mountains to the east coast was possible. Thus, it was assumed that advective venting occurred and a deep mountain ABL evolved. After around 1700 LT the humidity in the mountain ABL above the sea-breeze layer decreased again, which agreed with the cessation of topographic venting above the mountains due to decaying thermally driven circulations. Above the mountain ABL, large-scale advection of dry air with the westerly wind was visible, as described in Sect. 5.2.1.

On the west coast, westerly wind persisted (Fig. 5.17), which prevented advective venting of moist air from the island. No return flows of thermally driven circulations could establish, probably because of rather strong opposing westerly wind at around mountain-ridge height. This limited the mountain ABL depth on the west coast to the sea-breeze layer depth of about 500 m.

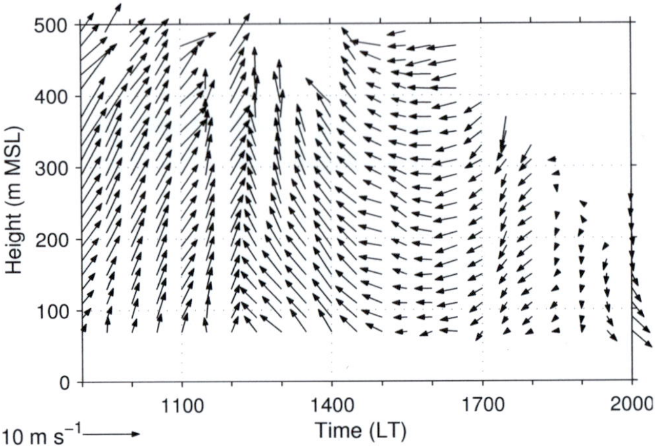

(a)

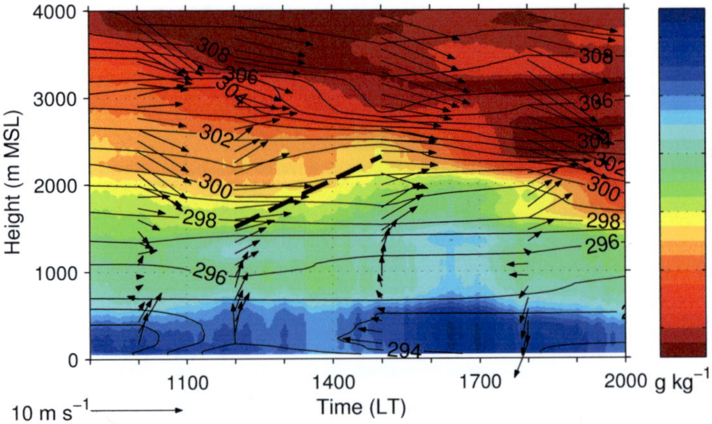

(b)

Figure 5.20.: Horizontal wind vector measured by sodar (a) and specific humidity (colour-coded), potential temperature (black isolines) and horizontal wind vector (arrows) measured by radiosondes (b) at San Giuliano on 02 October. In (b), specific humidity is interpolated using microwave radiometer and radiosonde data and potential temperature is temporally linearly interpolated between the radiosoundings and the mountain ABL height is indicated by the thick black dashed line.

The different evolutions of the mountain ABL on the coasts were also reflected in the diurnal cycle of IWV. At Ajaccio, without advective venting, the IWV evolution mostly expressed the large-scale advection of dry air (Fig. 5.18). The IWV at San Giuliano was likewise affected by the large-scale humidity decrease. However, it increased in the afternoon for a couple of hours due to the advective venting of moist air. Once the advective venting stopped, it decreased again.

5.2.3. Mountain ABL Evolution at Corte

The description of the mountain ABL evolution inland at Corte is divided into three time periods: before, during and after the event. These are time period I (about 0900 LT to 1600 LT), time period II (1600 LT to 1700 LT) and time period III (1700 LT to 2000 LT).

Time Period I The upvalley wind in the Tavignano Valley started just before 1100 LT, indicated by a wind shift to south-easterly directions near the surface at Corte (Fig. 5.21). At 1200 LT, the wind profile was characterized by a south-easterly to easterly flow of about 2 m s^{-1} in the lowest 1000 m AGL, i.e. far below the characteristic ridge height to the west, and by a westerly flow of about 8 m s^{-1} above (Fig. 5.22a). After 1330 LT, the near-surface wind speed accelerated to about 4 m s^{-1} (Fig. 5.21), which could be related to the arrival of the sea-breeze front at Corte from the east within the combined upvalley-wind and sea-breeze layer. The depth of the combined upvalley-wind and sea-breeze layer was about 1000 m at 1500 LT (Fig. 5.22a). In a zonal elevation scan performed with the wind lidar at Corte, this flow was clearly visible (Fig. 5.23a). Strong vertical wind shear between the south-easterly upvalley-wind layer and the westerly wind layer above resulted in a Richardson number R_i (Eq. 2.6) of less than the critical value $Ri_c = 0.25$ in the transition zone between the two wind regimes. Between 1200 and 1500 LT, the westerly wind weakened below about 1600 m AGL (Fig. 5.22a).

Horizontal humidity differences with higher humidity values above the ridges around Corte were present at around noon, which were likely to be related to topographic venting. Thus, advective venting of moist air at Corte could occur. Consequently, the

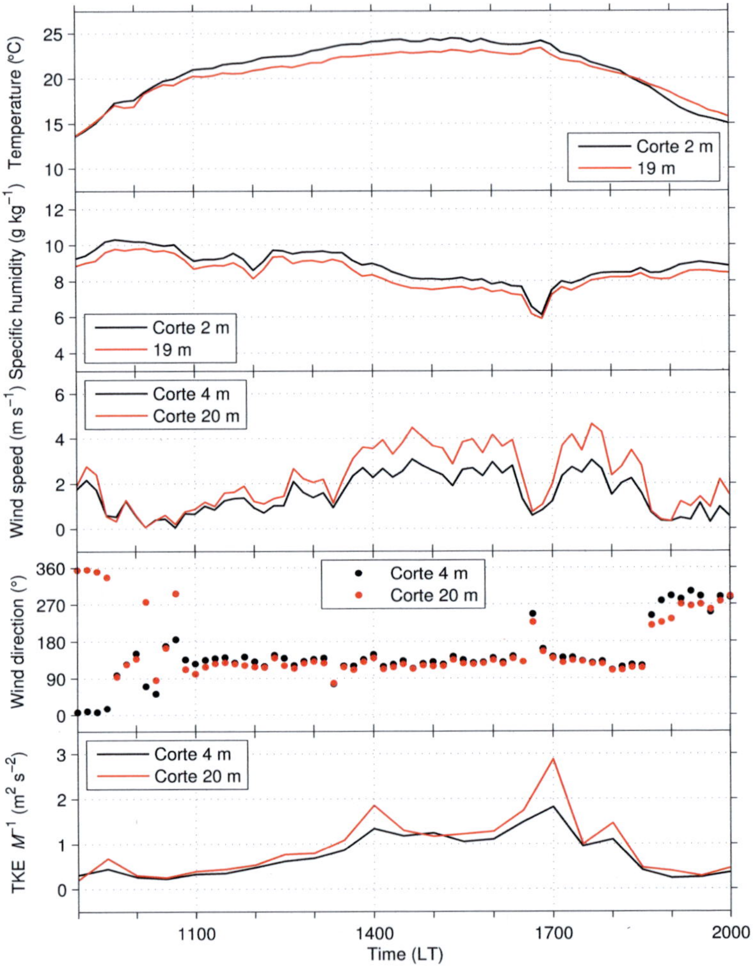

Figure 5.21.: Temperature, specific humidity, horizontal wind speed and wind direction and turbulent kinetic energy per unit mass (TKE M^{-1}) at Corte on 02 October.

increase of humidity above the upvalley-wind layer until 1500 LT (Fig. 5.22a), was attributed to the evolution of a deep mountain ABL up to about 1800 m AGL, i.e. it reached up to the mountain-ridge peaks. The mountain ABL height agreed well with the level of neutral buoyancy estimated from the near-surface potential temperature at Rusio and the radiosonde profile at Corte at 1500 LT.

Parallel to the evolution of the thermally driven circulations, a convection layer developed in the valley at Corte. Its height was determined using the combined vertical wind speed profiles from the two wind lidar systems (Sect. 4.1). The convection layer grew from about 100 m AGL at 1100 LT to about 900 m AGL at 1400 LT (Fig. 5.22b). Thus, its depth was lower than the valley-wind layer and mountain ABL depth. At 1500 LT, the convection-layer height dropped to about 600 m AGL. Because up- and downdraughts existed above the convection layer (Figs. 5.22b and c), σ_w^2 did not diminish but remained close to $\sigma_{w,\mathrm{th}}^2$ (Fig. 5.24a). At the same time, approximately height-constant profiles of potential temperature and specific humidity (Figs. 5.22a and 5.25a) established at Corte up to about 1600 m AGL (about 2000 m MSL), i.e. up to the characteristic ridge height in the west. Within this layer, the relative humidity increased with height and some shallow clouds with a cloud base at around 1800 m AGL occurred (Figs. 5.22b and c). Between 1500 and 1600 LT, the uncertainty of the convection-layer height was very large (>60 %). This could be ascribed to weak up- and downdraughts throughout the whole measurement range with valid data, which resulted in σ_w^2 ranging around $\sigma_{w,\mathrm{th}}^2$ (Fig. 5.24a), similar to the previous hour.

Time Period II After around 1600 LT, the conditions in the mountain ABL at Corte changed completely: dry and warm air gradually replaced the mountain ABL air, starting from top to bottom (Fig. 5.22a). Westerly wind of about 5 m s^{-1} being rather constant with height (Fig. 5.22a) and strong vertical upward motions exceeding 3 m s^{-1} (Figs. 5.22b and c) accompanied the intruding warm and dry air mass. This air mass was highly turbulent with σ_w^2 values of up to 4 m^2 s^{-2} and positive S values with very large errors existed (Fig. 5.24b). Due to the high σ_w^2, its value did not fall below $\sigma_{w,\mathrm{th}}^2$. Consequently, the highest measurement range gate with valid data at about 900 m AGL marked the convection-layer height (Figs. 5.22 and 5.24). Between about 1630 and 1700 LT, the dry and warm air reached the surface which was indicated by a temperature increase, specific humidity decrease, TKE increase, wind speed decrease and a shift to westerly wind (Fig. 5.21). In the same time period, the IWV at Corte decreased by about 5 kg m^{-2} (Fig. 5.18).

To determine the origin of the warm and dry air with a westerly wind component, profiles of specific humidity from radiosoundings at Corte were compared with upstream conditions at Ajaccio (Fig. 5.25a). As the atmosphere at Ajaccio was unaffected by advective venting (Sect. 5.2.2), it presumably resembled the conditions in the free atmosphere above the sea-breeze layer. Assuming a dry-adiabatic vertical displacement, specific humidity is a conserved quantity. Thus, the air mass replacing the mountain ABL air at Corte, originated from a layer between about 1300 and 2500 m MSL, i.e. from the free atmosphere below the elevated inversion at around 2500 m MSL. This layer roughly coincided with the flowing layer (Fig. 5.17).

In accordance with the radiosonde observations (Fig. 5.22a), the scanning microwave radiometer detected dry air gradually approaching Corte from the west. Figure 5.25b shows the spatio-temporal evolution of IWV measured during azimuth scans. At

---

Figure 5.22. *(facing page)*: Specific humidity (colour-coded), potential temperature (black isolines) and horizontal wind vector (arrows) measured by radiosondes (a); vertical wind speed measured by a wind lidar (b); and vertical wind speed measured by a cloud radar (colour coded) and specific humidity measured by radiosondes (black isolines) (c) at Corte on 02 October. Specific humidity is interpolated using microwave radiometer and radiosonde data and potential temperature is temporally linearly interpolated between the radiosoundings. The mountain ABL height is indicated by the thick black dashed line. In (b), the solid horizontal lines mark the convection-layer height and the thick dashed horizontal lines mark the lower and the crosses mark the upper boundaries of the uncertainty range, if present. The characteristic ridge height to the west (W) is indicated by the thin black dashed line. Dots mark the cloud base height measured by ceilometer in (b) and (c).

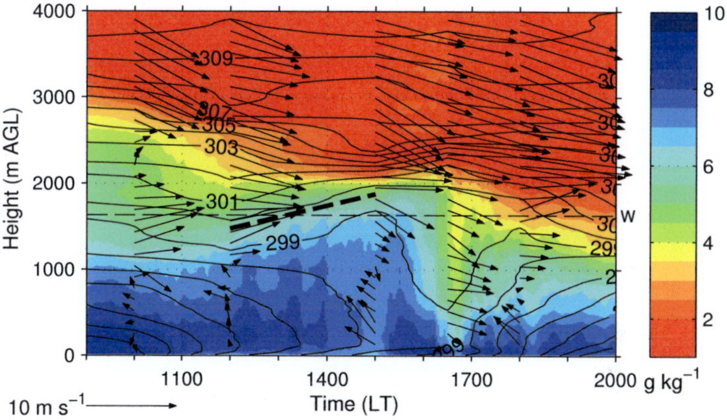
4000
3000
2000
1000
0
Height (m AGL)
309
307
305
303
301
299
W
1100
1400
1700
2000
Time (LT)
10 m s^{-1}
10
8
6
4
2
g kg^{-1}

(a)

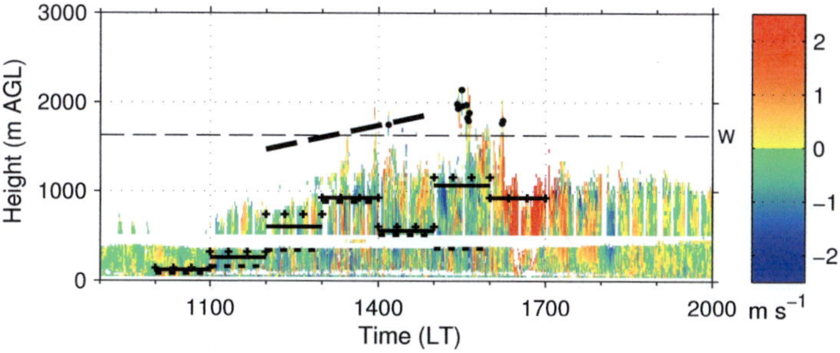
3000
2000
1000
0
Height (m AGL)
W
1100
1400
1700
2000
Time (LT)
2
1
0
−1
−2
m s^{-1}

(b)

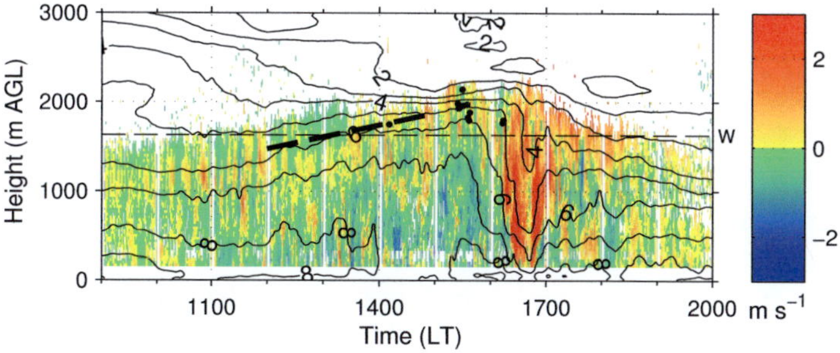
3000
2000
1000
0
Height (m AGL)
2
4
6
8
W
1100
1400
1700
2000
Time (LT)
2
0
−2
m s^{-1}

(c)

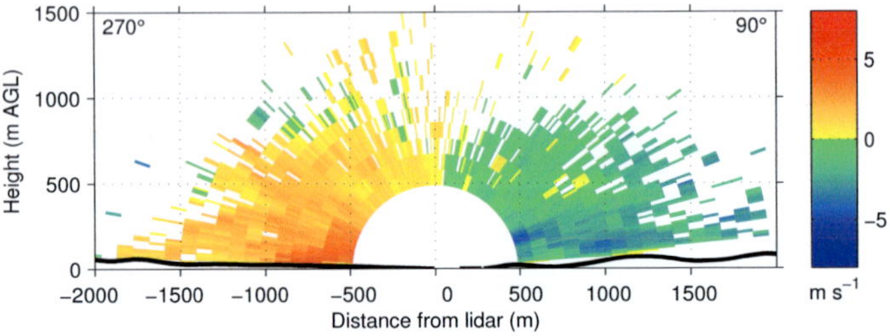

(a)

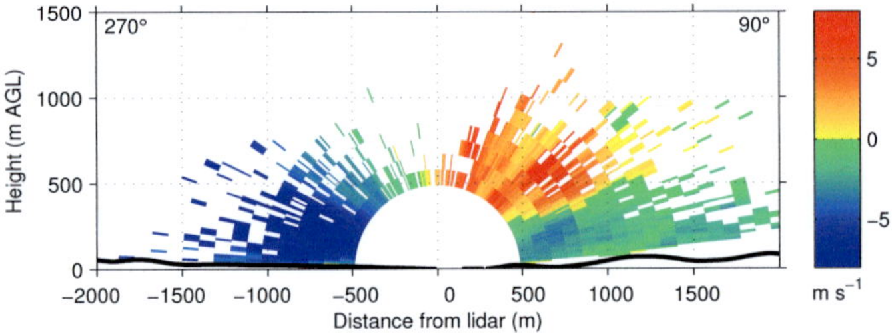

(b)

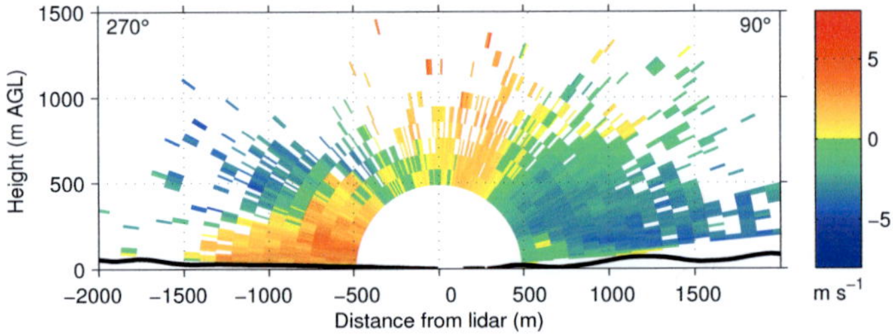

(c)

Figure 5.23.: Radial velocity measured by a wind lidar during a zonal elevation scan at 1600 LT (a), 1630 LT (b) and 1700 LT (c) at Corte on 02 October. Positive radial velocity indicates flow away from the lidar and negative radial velocity indicates flow towards the lidar. Azimuth directions of the scan are shown in the upper left and right corners of the figures and the black line denotes the orography along the scan.

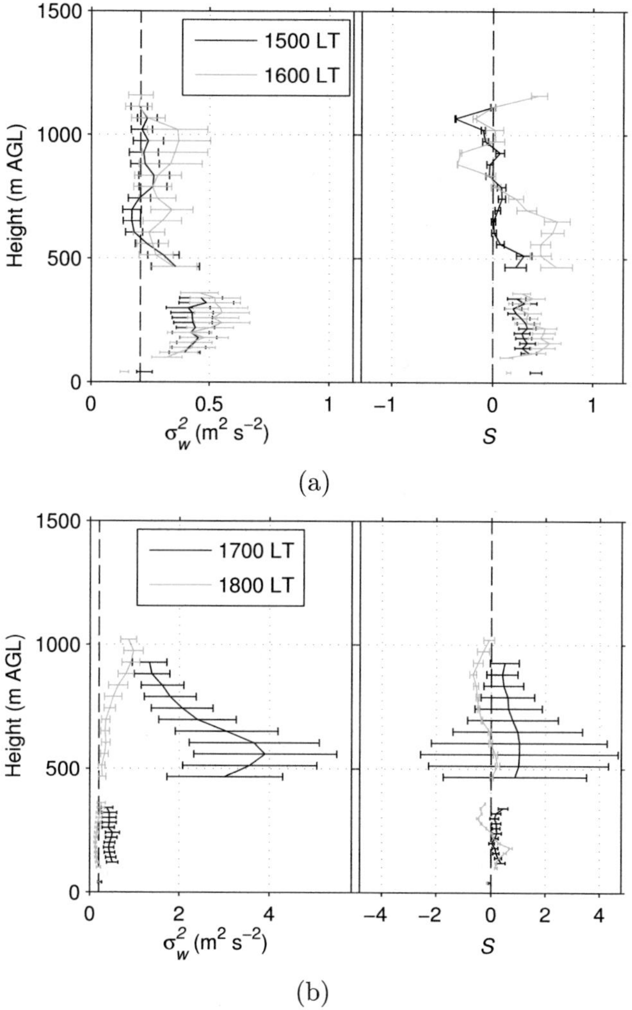

Figure 5.24.: Variance, σ_w^2, and skewness, S, of vertical wind speed measured by wind lidars at Corte at 1500 and 1600 LT (a) and 1700 and 1800 LT (b) on 02 October. The error bars show the total error and the vertical dashed black line indicates the used variance threshold and the zero skewness line, respectively. Note the different scalings of the x-axes.

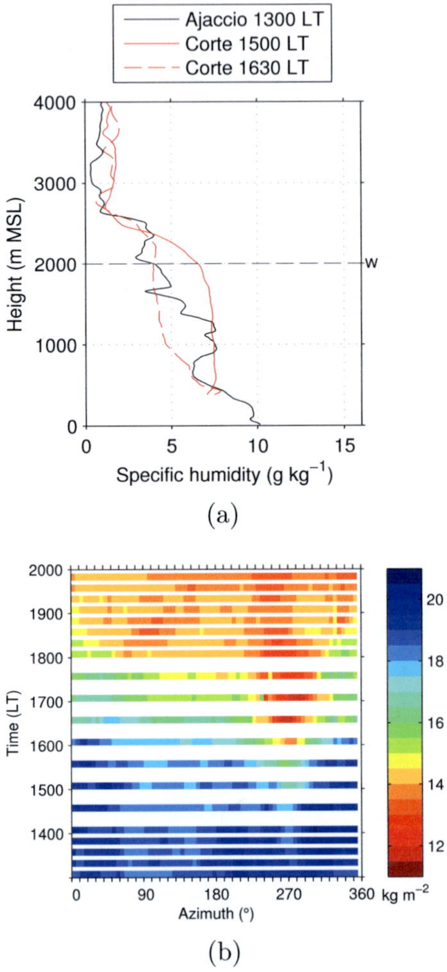

Figure 5.25.: Specific humidity measured by radiosondes at Ajaccio and Corte (a) and azimuth time plot of IWV for an elevation angle of 19.8° measured by a microwave radiometer at Corte (b) on 02 October. In (a), the characteristic ridge height to the west (W) is indicated by the black dashed line.

around 5.9 km horizontal distance from the radiometer, the height of the beam was about 2500 m MSL. Thus, first IWV differences could occur, when the dry air was within this distance from Corte. After around 1430 LT, the IWV in the west was slightly lower than in the other directions. These IWV differences increased and reached more than 5 kg m^{-2} at 1630 LT.

The strong changes of the mountain ABL characteristics around Corte were also reflected by the elevation scans of the wind lidar. In contrast to the elevation scan at 1600 LT, when the upvalley flow dominated at Corte (Fig. 5.23a), the elevation scan at 1630 LT revealed three distinct radial velocity pattern around Corte (Fig. 5.23b): (i) downward and strong westerly wind components to the west, (ii) an easterly wind component in a layer of about 300 m depth to the east and (iii) upward and westerly wind components above this layer. The layer with an easterly wind component represented the upvalley wind, which obviously still existed east of Corte. Between velocity patterns (i) and (ii), strong horizontal convergence occurred, which could cause the vertical upward motions (Figs. 5.22b, 5.22c and 5.23b).

Time Period III After around 1700 LT, the conditions at Corte changed again: near the surface, temperature and TKE decreased, specific humidity increased and the wind turned back to south-easterly directions and accelerated (Fig. 5.21). Simultaneously, the IWV increased (Fig. 5.18) and a moist upvalley-wind layer of almost 500 m depth prevailed at Corte again (Figs. 5.22a and 5.23c). Above the south-easterly upvalley wind, warm and dry air with a westerly wind component, upward motions and high σ_w^2 existed west of and at Corte within the next 30 minutes (Figs. 5.22, 5.23c and 5.24b). Above the upvalley-wind layer, the specific humidity remained lower than before the event (Fig. 5.22a), which was also reflected by lower IWV values (Figs. 5.18).

5.2.4. Interpretation of the Observations

On 02 October, a mountain ABL, which was deeper than the convection layer and in which potential temperature was mostly constant with height, developed downstream of the mountain ridge at Corte until the end of time period I (Fig. 5.22). While advective venting of moist air from the ridges likely contributed to the elevated humidity increase and evolution of the mountain ABL, another process was possibly responsible for the height-constant profiles of temperature and humidity within most of the mountain ABL: elevated vertical mixing above the convection layer could result from mechanical production of turbulence in layers where $Ri < Ri_c$. This assumption was supported by the subsequent observed reduction of wind shear (Fig. 5.22a) and by the non-diminishing σ_w^2 (Fig. 5.24a) in the relevant layers.

Once the layer with height-constant potential temperature extending up to the characteristic ridge height had evolved, a warm and dry air mass with a westerly wind component from the free atmosphere gradually intruded the downstream mountain ABL in the valley and eventually replaced the cool and moist upvalley-wind layer just west of Corte (time period II; Figs. 5.23b and 5.25b). Horizontal convergence resulted between the warm and dry air mass with a strong westerly wind component and the south-easterly upvalley wind inducing upward motions due to mass conservation (Figs. 5.22 and 5.23b). When the layer with horizontal convergence eventually reached the surface at Corte, which was indicated by a decrease in specific humidity and wind speed and by a temperature increase (Fig. 5.21), the cool and moist upvalley-wind layer had been pushed back completely down the valley to the east of Corte. The upward motions were accompanied by strong turbulence (Figs. 5.21 and 5.24b). Probably, turbulence was produced mechanically due to strong wind shear at the air mass boundaries and then advected with the mean and turbulent flow. The typical mountain ABL evolution with thermally driven circulations, topographic and advective venting and a convection layer was interrupted at Corte for about an hour. Because no interruption of the mountain ABL evolution on the coast could be observed, it was likely that the intrusion of free atmosphere air into the mountain ABL at Corte was spatially and temporally confined, i.e. it occurred on a horizontal scale of less than 10 km. However, it could not be excluded that the intrusion of free atmosphere air on 02 October lasted longer, but shifted out of the measurement range of the Corte station.

As stated in Sect. 2.4, several mechanisms may explain the intrusion of free atmosphere air and the replacement of mountain ABL air downstream of a mountain ridge: gravity waves formed above the mountains (Fig. 5.19b). A formation of large-amplitude lee waves is most likely when the wavelength of the gravity wave matches the terrain wavelength. On this day, the observed wavelength of the gravity wave was much shorter than the terrain wavelength and an amplification was unlikely. However, since the wavelength was measured above the intruding layer, it was not necessarily identical with the wavelength relevant for resonance.

For a downslope windstorm to develop, $Fr = 1$ has to apply in the flowing layer at the mountain ridge, i.e. the wind speed at the ridge has to equal the phase speed of the internal gravity wave of $c = 10$ m s^{-1}. An increase of the upstream wind speed from $U = 8$ m s^{-1} to $u_{\text{ridge}} = \frac{3}{2}U = 12$ m s^{-1} fulfilled this criterion approximately. The strong temperature inversion around ridge height (Fig. 5.22a) might additionally favour the evolution of a downslope windstorm. Although the upstream conditions would allow for a downslope windstorm to develop, typical characteristics of downslope windstorms were not observed, such as an acceleration of the flow on the downstream side compared to the conditions at the ridge or a hydraulic jump. If the latter two mechanisms were relevant on this day, a cessation of the downward transport could have been explained by a change in upstream conditions, modifying the wavelength or horizontal wind speed at the mountain ridge.

As the potential temperature was height-constant in the mountain ABL, turbulent downward transport of heat, humidity and momentum might have caused a gradual modification of the mountain ABL air mass. This mechanism seemed to be most likely on this day, because the replacement of the downstream mountain ABL air started only after a layer with height-constant potential temperature extending up to the mountain-ridge height had developed and strong wind shear existed due to opposing wind directions. Within the warm and dry air mass turbulence was high (Fig. 5.24b) and heat, humidity and momentum were rather uniformly distributed (Fig. 5.22a). When the vertical wind shear decreased and Ri increased, the turbulent transport might have been reduced leading to a return of the upvalley-wind layer and an upward shift of the convergence zone and associated upward motions (Figs. 5.22 and 5.23c).

5.3. The Inhibited Mountain ABL Evolution: Case Study of 09 October 2012

5.3.1. Large-Scale Conditions

On 09 October, the Corsican Island was located in the transition zone between a trough located over the eastern Mediterranean and a ridge with its axis over Spain and France. This resulted in strong north-westerly wind throughout the troposphere in the investigation area. The conditions upstream of the Corsican Island at 1300 LT were characterized by two distinct inversions at around 1500 m MSL and 3000 m MSL (Fig. 5.26). Below the lower inversion, a non-dimensional mountain height (Eq. 2.7) of $\hat{H} = 0.010\ \mathrm{s}^{-1} \cdot 1250\ \mathrm{m}\ \cdot (6\ \mathrm{m\ s}^{-1})^{-1} = 2.1$ was calculated using mean layer values, which indicated flow blocking. Shallow clouds formed

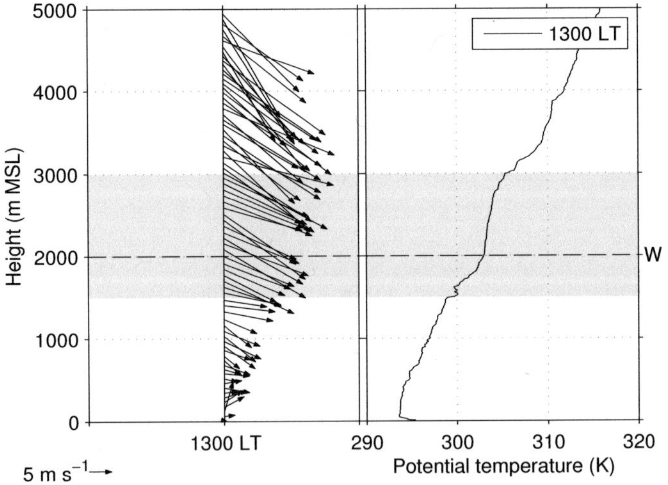

Figure 5.26.: Horizontal wind vectors and potential temperature measured by radiosonde at Ajaccio at 1300 LT on 09 October. Grey shading marks the flowing layer and the characteristic mountain-ridge height to the west of Corte (W) is indicated by the black dashed line.

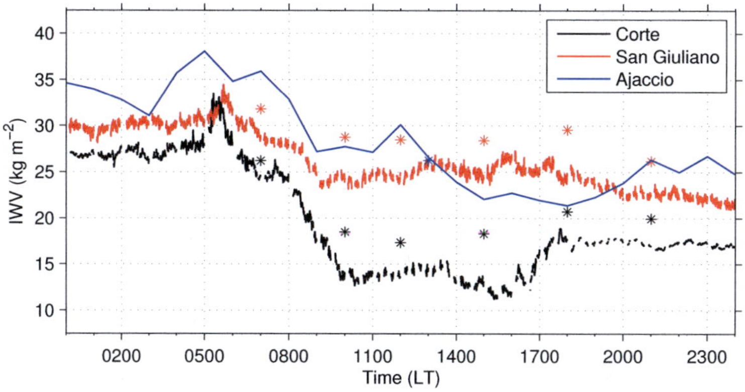

(a)

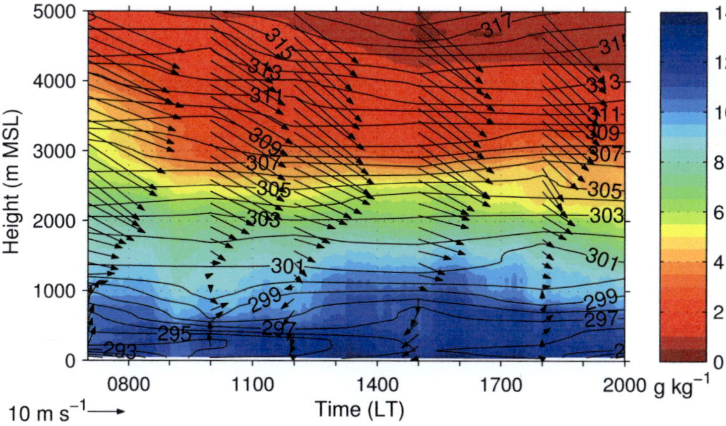

(b)

Figure 5.27.: IWV at Corte, San Giuliano and Ajaccio (a) and specific humidity (colour-coded), potential temperature (black isolines) and horizontal wind vector (arrows) measured by radiosondes at San Giuliano (b) on 09 October. In (a), the crosses indicate IWV calculated from radiosoundings and the solid lines mark IWV measured by radiometers at Corte and San Giuliano and GPS at Ajaccio. In (b), specific humidity is interpolated using microwave radiometer and radiosonde data and potential temperature is temporally linearly interpolated between the radiosoundings.

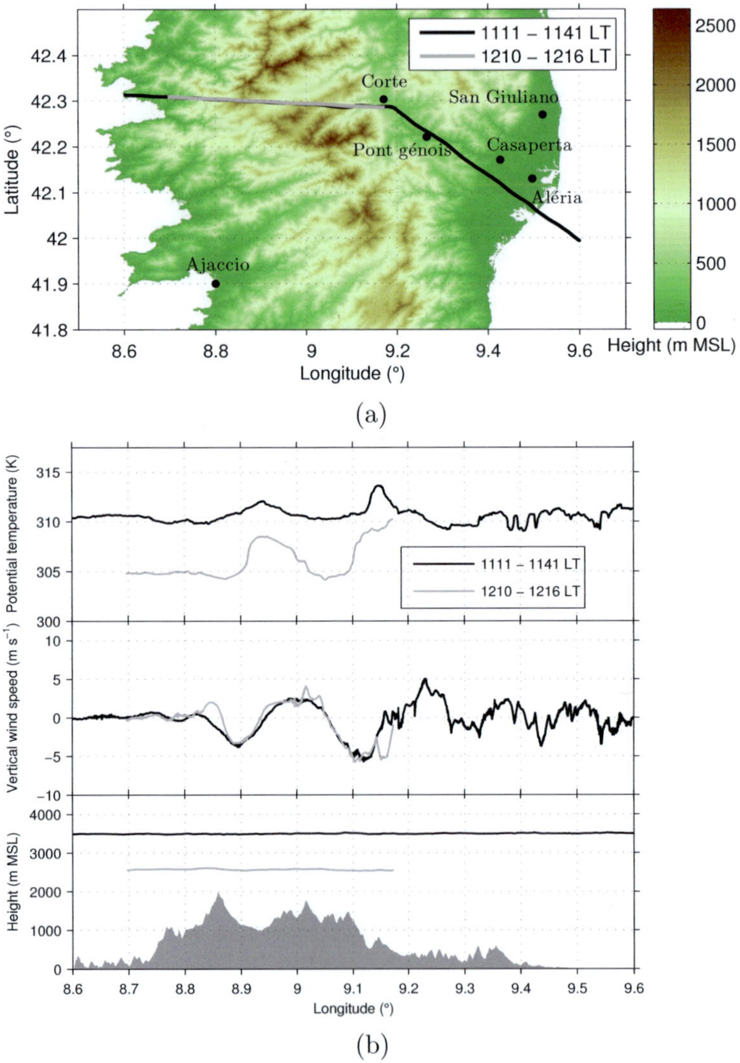

Figure 5.28.: Flight tracks with underlying orography (a) and potential temperature, vertical wind speed as well as flight level and orography along the flight tracks (b) on 09 October.

approximately in the middle of this layer, which slowly dissolved after 1300 LT. The layer confined by the two inversions was defined as flowing layer and was characterized by a mean north-westerly wind of $U \approx 17$ m s^{-1}, a mean potential temperature of $\Theta_{\mathrm{m}} \approx 303$ K and an upstream Froude number of hydraulic flow theory (Eq. 2.25) of $Fr = 17$ m s$^{-1}\cdot$ (17 m s^{-1})$^{-1} = 1$. During the second half of the day, the wind speed in the flowing layer weakened.

The IWV dropped by at least 5 kg m^{-2} at Corte, San Giuliano and Ajaccio between about 0600 and 0900 LT (Fig. 5.27a). This was caused by a specific humidity decrease of about 3 g kg^{-1} between about 2000 and 4000 m MSL due to large-scale advection, as was observed at San Giuliano (Fig. 5.27b). Although San Giuliano was located on the east coast and thus downstream of the island, the two elevated inversions characterising the upstream conditions (Fig. 5.26) were also visible, however at a somewhat lower height (Fig. 5.27b). The weakening of the flow in the flowing layer in the afternoon was also evident at San Giuliano.

Two zonal aircraft crossings roughly perpendicular to the mountain ridge at about 2500 m MSL, i.e. within the flowing layer, and at about 3500 m MSL around noon revealed the existence of gravity waves (Fig. 5.28), indicated by a 90° phase shift between the temperature and vertical wind-speed waves (Sect. 2.4.2). Above the mountain ridge, oscillations of vertical wind speed with an amplitude of up to 5 m s^{-1} and a wavelength of about 20 km occurred, which agreed well with the terrain wavelength. Upstream, i.e. to the west of the mountain ridge, the fluctuations were small. Downstream of the mountains, when the aircraft followed the axis of the Tavignano Valley, the amplitude of the vertical wind-speed wave was reduced (Fig. 5.28) and TKE was lower.

5.3.2. Mountain ABL Evolution over the Corsican Island

Similar to case study II, this analysis is split into three time periods: time period I (0700 to 1000 LT) describes the onset of the event in the Tavignano Valley; time period II (1000 to 1430 LT) its duration; and time period III (1430 to 2000 LT) its cessation.

Time period I At 0700 LT, a strong nocturnal surface inversion below about 700 m AGL existed at Corte (Fig. 5.29a). Within this

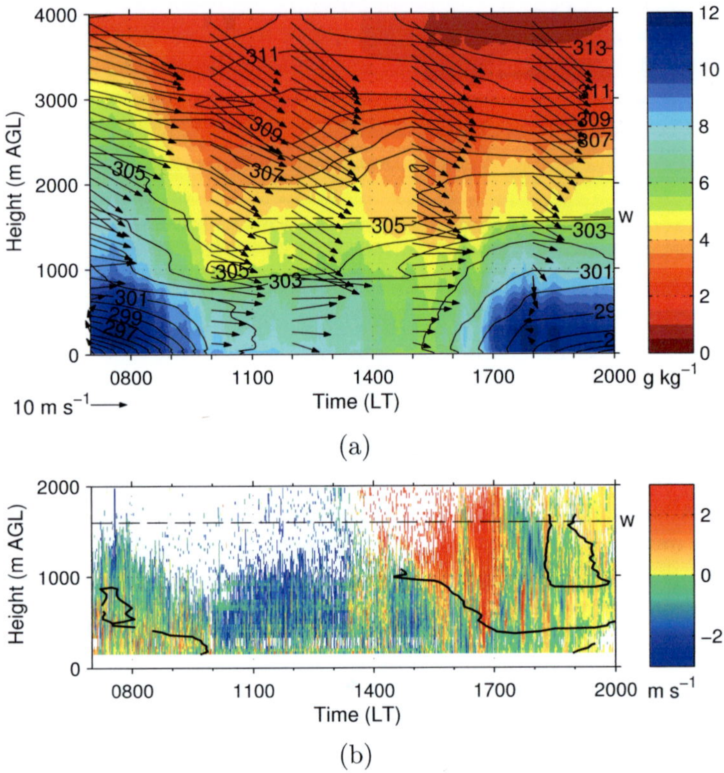

Figure 5.29.: Specific humidity (colour-coded), potential temperature (black isolines) and horizontal wind vector (arrows) measured by radiosondes (a) and vertical wind speed (colour-coded) and 30-minute averaged vertical wind speed (0 m s^{-1} isoline) measured by cloud radar (b) at Corte on 09 October. In (a), specific humidity is interpolated using microwave radiometer and radiosonde data and potential temperature is temporally linearly interpolated between the radiosoundings. The characteristic ridge height to the west (W) is indicated by the black dashed line.

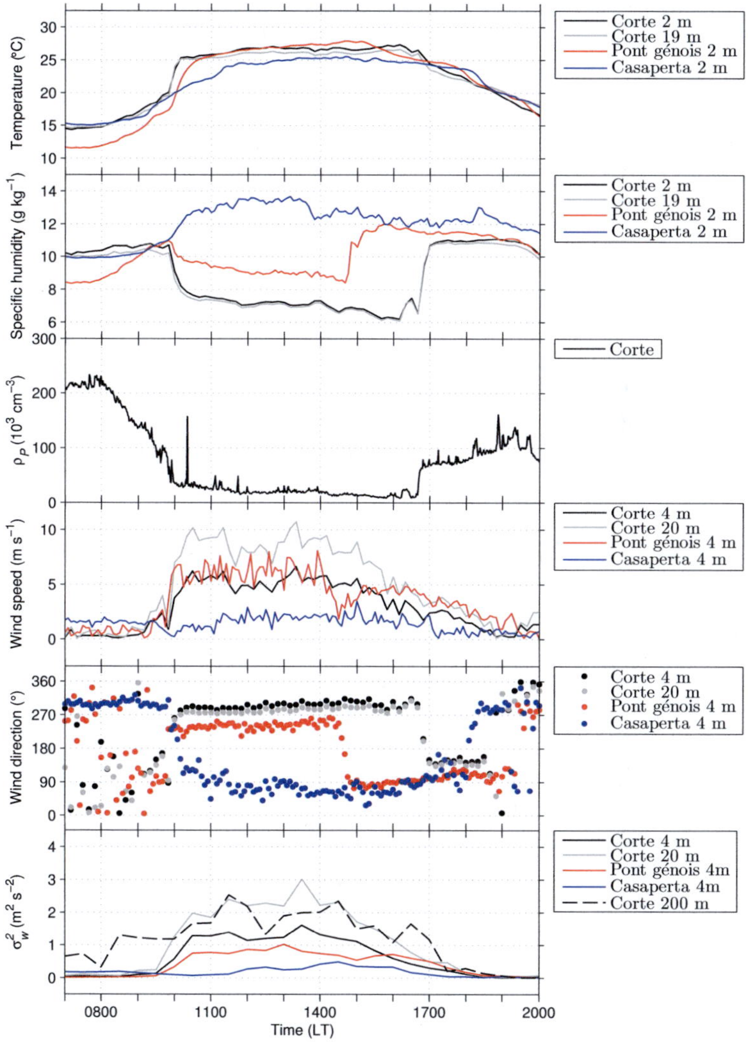

Figure 5.30.: Temperature, specific humidity, near-surface particle number density, ρ_P, horizontal wind speed, horizontal wind direction and vertical wind speed variance, σ_w^2, at different stations and at different heights on 09 October. σ_w^2 was calculated from vertical wind-speed data measured by flux towers and wind lidar for 30-minute intervals.

inversion, a weak south-easterly flow prevailed. Above the inversion, the stable stratification was weaker and the horizontal wind turned to large-scale north-westerly direction and rapidly increased with height to more than 10 m s^{-1}. Above the mountain ridge peaks, the wind speed was approximately as high as on the upstream and far downstream side (Figs. 5.26, 5.27b and 5.29a). The layers with strong vertical wind shear were associated with Richardson numbers (Eq. 2.6) of $Ri < Ri_c$. Strong mean subsidence above the surface inversion decelerated when it penetrated into the inversion (Fig. 5.29b). The layer with subsidence was associated with a lower specific humidity and higher potential temperature (Fig. 5.29). Aerosol-free air gradually replaced the valley air mass, as obvious from the cloud-radar signal (Fig. 5.29b).

Once the nocturnal surface inversion was eroded at 1000 LT, the mean subsidence almost reached the ground (Fig. 5.29). A layer in which potential temperature and specific humidity were about constant with height existed up to about 1000 m AGL topped by an elevated inversion. This inversion layer was accompanied by a local wind-speed maximum of about 15 m s^{-1}. Parallel to the erosion of the nocturnal surface inversion, a sudden near-surface temperature increase ($>$ 5 K), specific humidity decrease ($<$ -3 g kg^{-1}) and particle number density decrease occurred at Corte (Figs. 5.30). The wind turned to mainly westerly direction and increased to more than 5 m s^{-1} at 4 m AGL and to more than 10 m s^{-1} at 20 m AGL. The variance of the vertical wind speed strongly increased at 4 and 20 m AGL. At 200 m AGL, the increase took place a little earlier. The IWV at Corte dropped by more than 12 kg m^{-2} between 0800 and 1000 LT, which was a significantly stronger decrease than at San Giuliano and Ajaccio (Fig. 5.27), where the IWV decrease was mainly caused by large-scale advection of dry air (Sect. 5.3.1). Similar significant changes in near-surface conditions were simultaneously observed at Pont génois (Figs. 5.30), about 15 km down the valley, although the specific humidity decrease was about 1.5 g kg^{-1} lower than at Corte, the σ_w^2 increase was less pronounced and the wind shifted to south-westerly instead of westerly.

Further down the valley at Casaperta the near-surface conditions showed a mostly undisturbed evolution of an upvalley wind between about 1000 and 1800 LT (Fig. 5.30). On the east coast a sea breeze existed during the day (Fig. 5.27b). Its depth reached more than 500 m (Fig. 5.27b), which was about typical in October (Sect. 5.2.2). Above the sea breeze up to about 1500 m MSL, an increase of humidity of more than 1 g kg^{-1} occurred until 1300 LT and remained high during the afternoon. As the moistening was not associated with a westerly component, i.e. flow from the mountain ridges, it was not assignable to advective venting and the evolution of a deep mountain ABL on the east coast.

Time period II The above conditions remained about the same until the afternoon. The layer with height-constant potential temperature and specific humidity and the elevated inversion with a horizontal wind-speed maximum reaching even more than 20 m s^{-1} persisted (Fig. 5.29a) and were accompanied by $Ri < Ri_c$. Within these two layers, potential temperature, specific humidity, wind speed and wind direction varied only slightly with time and the particle number density and IWV remained low (Figs. 5.27a, 5.29a and 5.30). High σ_w^2 values were observed at 4, 20 and 200 m AGL (Fig. 5.30). As aerosol was removed from the atmosphere on this day, σ_w^2 could not be calculated for higher altitudes from wind-

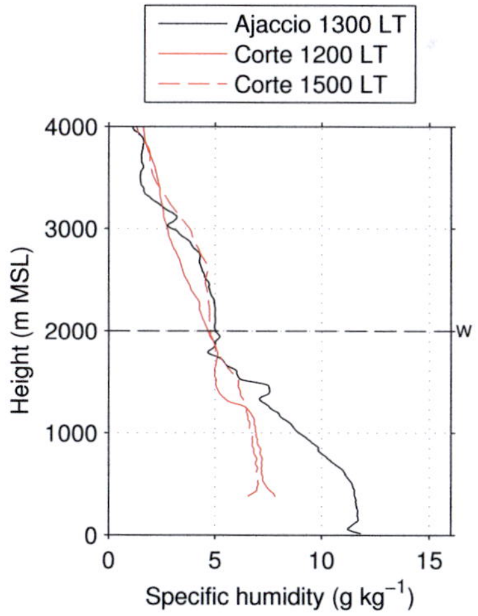

Figure 5.31.: Specific humidity measured by radiosondes at Ajaccio and Corte on 09 October. The characteristic ridge height to the west (W) is indicated by the black dashed line.

lidar data. For the same reason, no convection-layer height could be determined with wind-lidar data. However, enough scatterers were present below the elevated inversion to allow for vertical wind speed measurements with the cloud radar. In this layer, no surface-based updraughts were visible (Fig. 5.29b), which typically accompany the evolution of a convection layer. Instead, persistent strong mean subsidence occurred, but with embedded turbulence as indicated by high σ_w^2 values (Fig. 5.30).

The origin of the warm and dry air mass with a westerly wind component close to the surface at Corte was not as clear as for case study II. Comparison of the specific humidity profiles at Corte with the upstream conditions at Ajaccio indicated that the air mass originated from the flowing layer between about 1500 and 3000 m MSL (Fig. 5.31), i.e. from the free atmosphere. The air above that layer was much drier and the air below was moister than observed at Corte.

Time period III While the near-surface conditions at Corte remained about constant until about 1630 LT, they changed significantly at Pont génois at around 1430 LT (Fig. 5.30): the temperature decreased by about 3 K, specific humidity increased by about 3 g kg^{-1} and the wind decelerated and turned to easterly upvalley direction.

At around 1530 LT and 1630 LT, the subsidence in about the lower 1000 m AGL at Corte was interrupted by two events with strong upwinds, which were associated with a humidity increase (Fig. 5.29). After the second upwind event, the pre-upwind event conditions did not return and rapid changes occurred near the surface at Corte (Fig. 5.30): temperature and σ_w^2 decreased, specific humidity and particle number density increased and the wind turned to south-easterly upvalley direction. Simultaneously, the IWV increased (Fig. 5.27a). The humidity values within about the lower 1000 m AGL re-attained the same level as before 1000 LT (Figs. 5.29a and 5.30). Due to the drier air above about 1000 m AGL, the IWV did not reach pre-event values (Fig. 5.27a). At 1800 LT, a typical upvalley-wind system of about 800 m depth existed and at 1900 LT, a nocturnal downvalley-wind system established (Figs. 5.29a and 5.30).

5.3.3. Interpretation of the Observations

On 09 October, the observations indicated that the background flow dominated the conditions in a major part of the Tavignano Valley. The penetration depth of the warm and dry air mass with a westerly wind component depended on the downstream stratification, as the subsiding air did not intrude into the nocturnal surface inversion (time period I; Figs. 5.29 and 5.30). The sensible surface heat flux as well as turbulent erosion at the top of the inversion due to strong wind shear contributed to the removal of the nocturnal surface inversion in the morning. Once the surface inversion was eroded and a neutrally stratified layer established, the warm, dry and aerosol-free air mass with a westerly wind component from above reached the ground (Fig. 5.30). As Corte is located a few kilometres downstream of the mountains, the downward motion (Fig. 5.29b) obviously occurred in quite some horizontal distance from the mountain ridge. The intrusion of warm and dry air mass with a westerly wind component also affected the conditions far down the Tavignano Valley, as was confirmed by the measurements at Pont génois. The surface observations at Corte and Pont génois showed a strong similarity to foehn events, i.e. an increase of temperature and wind speed and a drop in humidity occurred simultaneously within minutes (e.g. Richner and Hächler, 2013). Further towards the valley exit and on the east coast typical thermally driven circulations evolved.

The intruding air mass inhibited the evolution of a mountain ABL and prevented the upvalley-wind layer from reaching Corte and Pont génois until the afternoon (time period II; Fig. 5.30). This allowed warm and dry air and a strong westerly wind component to be measured at Corte and Pont génois for most of the day. A convergence zone developed between the westerly flow and the south-easterly upvalley-wind layer and existed somewhere between Pont génois and Casaperta.

Eventually, the upstream flow decelerated in the afternoon and allowed the upvalley wind and the associated convergence zone to push forward (time period III). The cool and moist upvalley-wind layer first replaced the warm and dry air with a westerly wind component at Pont génois and then moved further upvalley towards Corte (Fig. 5.30). The air within the upwinds, accompanying the convergence zone, was likely mixed from the two converging air masses and was therefore moister than the intruding air mass (Fig. 5.29). Between 1530 and 1700 LT, this convergence zone billowed around Corte. After that, the intruding air mass and the convergence zone were shifted upvalley of Corte, allowing the upvalley-wind layer to determine the conditions at Corte. Thus, a mountain ABL finally established at Corte. As no elevated humidity increase due to advective venting was observed, its depth was confined to the upvalley-wind layer.

To explain the observations in the downstream atmosphere, the possible mechanisms specified in Sect. 2.4.2 are consulted. The aircraft observations revealed the evolution of persistent large-amplitude lee waves (Fig. 5.28b). As the wavelength of the lee waves approximately matched the terrain wavelength, an amplification was possible.

According to the hydraulic flow theory, the wind speed in the flowing layer accelerated from $U = 17$ m s^{-1} to $u_{\mathrm{ridge}} = \frac{3}{2}U = 26$ m s^{-1}, which far exceeded the phase speed of the internal gravity wave of $c = 18$ m s^{-1}. Assuming that these estimations were done for a meaningful layer and that the hydraulic flow theory applies, this would not favour the evolution of a downslope windstorm. Additionally, no increase of wind speed in the intruding air mass on the downstream side (<20 m s^{-1}) compared to the wind speed at the mountain ridge (≈ 26 m s^{-1}) was observed. However, the abrupt changes of near-surface conditions, the persistency of the downward transport and the far extension along the valley axis would support this mechanism as an explanation.

With the evolution of a layer where the potential temperature was approximately constant with height, additional turbulent transport of heat, humidity and momentum to the ground might have occurred. To maintain the height-constant profiles of almost 1000 m depth (Fig. 5.29), turbulent mixing was necessary, which could have been produced by buoyancy or vertical wind shear around the inversion layer at around 1000 m AGL. This mechanism was favoured by the observed high turbulence below the inversion (Figs. 5.30).

It was assumed that in the morning, when the atmosphere was stably stratified, the most likely mechanisms were the evolution of large-amplitude lee waves or a downslope windstorm, even though hydraulic flow theory revealed different results. During the day, at least parts of the downward transport in lower layers were probably caused by turbulence, as this was the only mechanism to maintain height-constant profiles of potential temperature and specific humidity.

5.4. Discussion

The mountain ABL in the centre of the Corsican Island at Corte underwent a distinct diurnal evolution, unless it was impacted by the background flow. From the existing measurements it was hard to predict whether the evolution of the mountain ABL at Corte would be undisturbed, interrupted or even inhibited, as all three case studies were initially characterized by clear skys and high surface sensible heat fluxes. Nevertheless, the diurnal evolution of the mountain ABL varied. It is hypothesized that a complex combination of upstream conditions, stratification and surface energy input determined the different developments.

Without an impact of the background flow, a mountain ABL evolved as a combination of convection and mesoscale processes (Sect. 5.1). The case study revealed experimental evidence of multi-scale processes affecting the characteristics of the mountain ABL in a valley. These included e.g. a convection layer; strong, long-lasting subsidence; topographic and advective venting; elevated up- and downdraughts; and vertical coupling of surface-based convective cells to elevated updraugths. The convection layer in the valley centre was usually lower than the mountain ABL. Its evolution obviously depended on various factors, like the surface sensible heat flux, stratification in the valley atmosphere and the strengths and depths of the upvalley- and upslope-wind systems. On the presented day, the convection-layer depth showed a diurnal cycle, although it was possibly reduced by mean subsidence and enhanced by vertical coupling with elevated updraughts. On other days with undisturbed mountain ABL evolutions, especially in September and October (Figs. B.2 to B.5 in the appendix), only very shallow or no continuous convection layers formed in the valley centre when applying the detection algorithm introduced in Sect. 4.1. The mean, long-lasting subsidence in the valley centre was generally observed on every day with an undisturbed mountain ABL evolution, although its duration and intensity varied somewhat. Thus, it is likely to be a typical characteristic of the mountain ABL and to regularly affect the convection-layer growth in the valley. The analysis of other days further revealed that the cell coupling between the convection layer and an elevated layer in the valley centre was not observed every day. It is hypothesized that it depended on the vertical extent of the convection layer and on the stratification above: if the convection layer was too low, vertical distance between the two layers might be too large to overcome. If the stability in the elevated layer was too high, the vertical motions might be damped before they could reach the valley centre. As this previously unreported process likely represented an effective transport mechanism, provided that it occurred, it should be further explored.

The layers with height-constant potential temperature and specific humidity, which usually represent the well-mixed layer of a classical CBL over homogeneous terrain, neither resembled the convection layer nor necessarily the mountain ABL in the Tavignano Valley. This is because other processes than convection, such as mesoscale advection, significantly contributed to the profiles of mean variables measured by radiosondes. Even when height-constant profiles existed, their tops might not be associated with strong gradients, which might cause problems when applying standard methods to determine the CBL height to data measured over complex terrain. Because of the behaviour of the atmospheric stratification, it is questionable if the detection of a classical CBL over complex terrain makes sense at all. It is rather suggested that a mountain ABL and a convection layer might be more meaningful layers to describe the impact of the Earth's surface on the atmosphere over complex terrain. It would not have been possible to determine the convection-layer height without wind lidar measurements or the mountain ABL height without frequent radiosoundings. Therefore, it is recommended to be cautious when characterizing the atmospheric structure over complex terrain on the basis of a single radiosonde and without knowledge of turbulence profiles.

The evolution of a mountain ABL may also be treated as a handover process between the CBL and the free atmosphere. In this case, the CBL only includes layers adjacent to the surface that are directly influenced by buoyancy-driven surface forcing; the upper branches of thermally driven circulations and topographic and advective venting are not considered to occur in the boundary layer and the above problem concerning the CBL determination arises. For example, Kossmann et al. (1999) or Kalthoff et al. (2000) have used this approach to describe the transport of water vapour and trace gases from the CBL to the free atmosphere. This concept may be especially justified when elevated layers downstream of the mountainous area are examined, which result from advective venting and the earlier transport of water vapour into the free atmosphere by topographic venting. Depending on the environment and on the downstream distance, the response time scales may be rather long, the diurnal cycle may be weak and the large-scale flow may become more dominating. Consequently, the direct relation between the elevated moist layers, thermally driven circulations and the CBL ceases. In particular when the transported water vapour affects layers much higher than the local downstream CBL, i.e. no diurnal changes are observed in a layer between the CBL and the elevated moist layer, it might be more adequate to refer to handover processes instead of a mountain ABL. On the other hand, when analysing the diurnal evolution of atmospheric conditions over mountainous terrain, it may be very difficult to distinguish the layers dominated by convection or thermally driven circulations, especially when the observations are limited. Thus, it may be not reasonable to separately treat the lower and upper branches of thermally driven circulations, the CBL and layers with advective and topographic venting, but rather investigate the layer that undergoes a diurnal variation as a whole, according to the concept of the mountain ABL. As the elevated layers on the downstream coast of the Corsican Island showed a diurnal cycle and revealed a response time scale of a few hours, they were considered to be part of the mountain ABL.

The typical diurnal evolution of the mountain ABL differed at Corte, when the background wind impacted the mountain ABL and warm and dry air with a westerly wind component intruded the atmosphere downstream of the mountain ridge. On a day with very strong upstream wind from westerly directions, a mountain ABL evolution was inhibited over a large part of the downstream valley, lasting for several hours (Sect. 5.3). Only far downstream of the mountain ridge thermally driven circulations and a mountain ABL developed. An interruption of the mountain ABL evolution and an intrusion of warm and dry air from the free atmosphere into the downstream mountain ABL even occurred when the upstream wind was not particularly strong (Sect. 5.2). In this case, the intrusion existed on a smaller scale, i.e. few tenth of kilometres, and lasted for a limited time period. This enhances the probability that such intrusion events are not resolved by the measurement network and it can not be excluded that they do occur more often.

It is hypothesized that the conditions measured at Corte depended on the location of the convergence zone between the warm and dry air mass with a westerly wind component and the cool and moist upvalley-wind layer. While the convergence zone billowed around Corte during the event on 02 October and produced upward motions at the station (Fig. 5.32a), it was located much further down the valley between Pont génois and Casaperta on 09 October and therefore allowed warm and dry air and a westerly wind component to be measured for most of the day at Corte and Pont génois (Fig. 5.32b). The location of the convergence zone further away from the mountain ridge on 09 October could have been caused by a stronger westerly wind component in the air mass that replaced the mountain ABL air, as the strength of the upvalley wind was about the same on both days.

One main difference between both days was that the upward motions were accompanied by dry and warm air on 02 October and by moist and cool air on 09 October compared to the conditions previously existing at Corte. It was assumed that the air within the upwinds was a combination of two air masses, the warm and dry intruding air mass and the cool and moist upvalley-wind layer. Thus, the air appeared to be warm and dry when Corte was located in the upvalley-wind layer before the onset of the upwinds, while the upwinds were associated with cooler and moister air when the intruding air mass prevailed at Corte beforehand.

To explain the downward transport of a warm and dry air mass with a westerly wind component, several possible mechanisms were investigated, i.e. turbulent transport, large-amplitude lee waves or downslope windstorms. In agreement with the different characteristics observed, it is assumed that the dominating mechanism on 02 October was turbulent transport, while on 09 October, lee waves and a downslope windstorm were more important. This could explain the much shorter duration and smaller horizontal scale of the downward transport on 02 October. The suggested transport mechanisms required a deep layer with height-constant potential

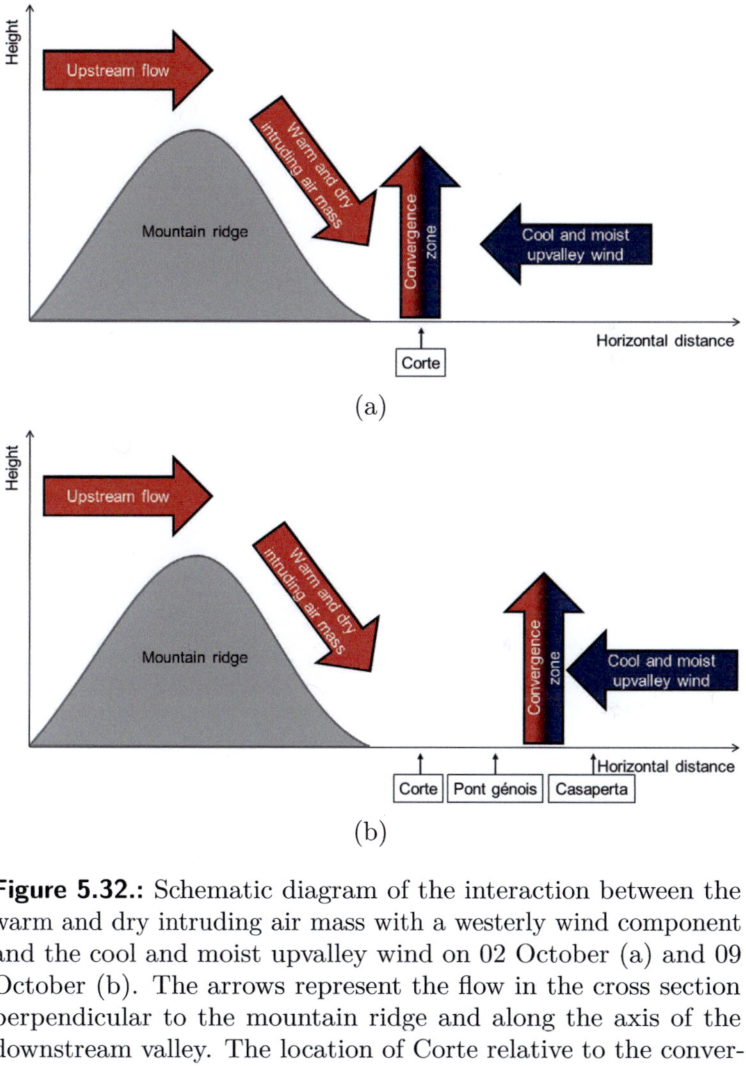

Figure 5.32.: Schematic diagram of the interaction between the warm and dry intruding air mass with a westerly wind component and the cool and moist upvalley wind on 02 October (a) and 09 October (b). The arrows represent the flow in the cross section perpendicular to the mountain ridge and along the axis of the downstream valley. The location of Corte relative to the convergence zone during the investigated event is indicated. In (b), the locations of Pont génois and Casaperta are additionally marked.

temperature on the downstream side, strong wind shear or lee waves or a combination of these both. The occurrence of these characteristics in the atmosphere is unexceptional, which increases the likelihood of an intrusion of air from the free atmosphere.

6. Effect of the Mountain ABL on Water Vapour and Atmospheric Stratification

In Sect. 5, boundary-layer processes were analysed in detail for days with different mountain ABL evolutions. The conducted case studies revealed that the evolution of the mountain ABL was closely related to the water-vapour distribution over the Corsican Island. To assess the representativity of the found relation and to further investigate the mesoscale water-vapour variability, the spatio-temporal behaviour of water vapour was investigated for the whole island over a three-month period and for additional case studies (Sect. 6.1).

Secondly, the connection between the mountain ABL and the evolution of isolated deep convection over the Corsican Island was examined. By means of convection-related parameters, the influence of the mountain ABL on water vapour and atmospheric stratification, which are decisive for the evolution of deep convection, was analysed and the mesoscale variability of the convection-related parameters in connection with the evolution of deep convection was studied (Sect. 6.2).

Like in the previous section, only the day and month are given in the date information as the analysis in the following section is based on data from the year 2012 only.

6.1. Variability of Water Vapour over the Corsican Island

The temporal evolution of IWV at Corte in the centre, Solenzara on the east coast and Ajaccio on the west coast from August to November is shown in Fig. 6.1. Days were selected when the temporal IWV evolution was presumably primarily controlled by

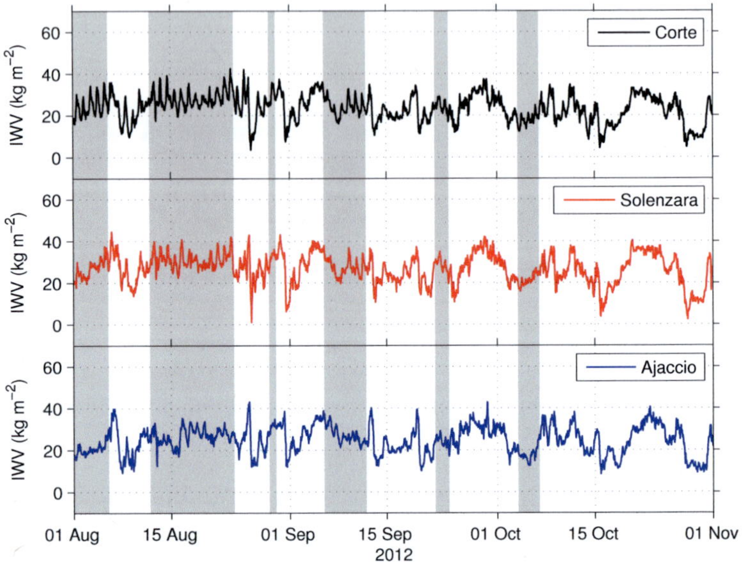

Figure 6.1.: IWV derived from GPS measurements at Corte, Solenzara and Ajaccio. The grey vertical bars mark the days used to analyse the spatial IWV distribution over the Corsican Island.

processes in the mountain ABL and not by large-scale air mass changes. Air mass changes were identified by a nearly simultaneous strong change of IWV at all three sites. After applying this criterion, 35 days remained, which are marked by the grey vertical bars in Fig. 6.1. These days mainly occurred in August and the first half of September, because large-scale systems more frequently affected the Corsican Island in the second half of the measurement period. This agrees with the climatology for heavy precipitation events for the western Mediterranean basin (Ricard et al., 2012). The selection still includes days when isolated deep convection developed over the Corsican Island (23 August, 25 August, 06 September and 09 September). On these days, the deep convective cells were locally initiated by processes related to the orography and not by large-scale lifting and were also not embedded in large-scale systems.

The IWV value measured by a ground-based instrument generally depends on the station height, because the volume of a vertical air column above the station decreases with increasing station height. The IWV is further influenced by the atmospheric conditions over the station, e.g. to what extent the atmosphere above the station is affected by a mountain ABL. To investigate this, a reference atmosphere is defined which is assumed to be unaffected by the evolution of a deep mountain ABL. Even if the atmosphere above a station is not influenced by a deep mountain ABL, at least evapotranspiration from the surface normally contributes to an IWV increase during the day. In Sect. 5.1, this contribution was estimated to be in the order of 1 kg m^{-2} under typical conditions for the Corsican Island. To investigate the atmospheric conditions over a station in relation to the reference atmosphere, the water-vapour density of the atmosphere above the station was compared with that in the reference atmosphere at the same level. Because the flow field measured by radiosondes at

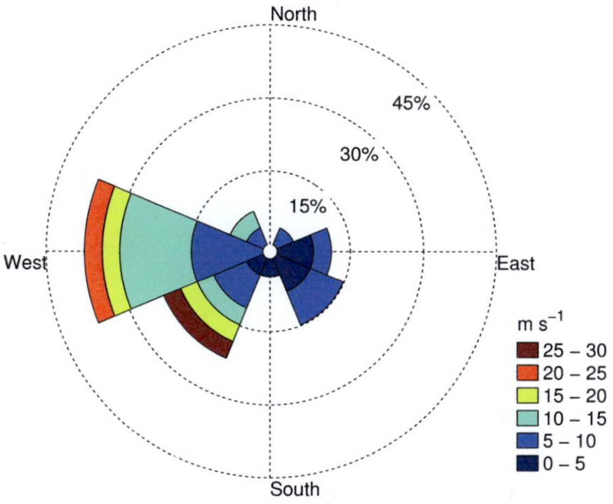

Figure 6.2.: Frequency of wind speed and wind direction measured by radiosondes at 700 hPa at Ajaccio around noon for the days marked by the grey vertical bars in Fig. 6.1.

Ajaccio revealed mainly westerly winds at 700 hPa at noon during the selected days (Fig. 6.2), the profiles at Ajaccio were measured mostly upstream of the island. Thus, the profiles were assumed to be mainly unaffected by the orography and to reflect the conditions in the reference atmosphere. The profile of IWV was calculated for each radiosounding from the water-vapour density profile in the air column above:

$$\mathrm{IWV}(z) = \int_z^{h_t} \rho_v(z')\mathrm{d}z'. \tag{6.1}$$

This allows to use the profile for data interpretation: when the IWV at a GPS station is significantly higher than the IWV value at the same height of the reference profile, this indicates that the atmosphere above the station is influenced by a deep mountain ABL. On the other hand, when the IWV value at a GPS station is close to the IWV value in the reference atmosphere at the respective level, the station is mainly unaffected by the evolution of a deep mountain ABL.

To obtain information about the spatial distribution of water vapour over the Corsican Island, IWV derived from the permanent GPS network was used. A total of 16 stations was rather uniformly distributed over the island (Fig. 6.3). Figure 6.4a shows the IWV profile measured by radiosondes at Ajaccio at noon averaged over all selected days and the IWV values from GPS stations between 1300 and 1400 LT, i.e. in the early afternoon, for the same days. As the radiosondes were launched at 1300 LT, the IWV values were valid for roughly the same time interval. In the reference atmosphere, IWV decreased by roughly 1 kg m^{-2} $(100\ \mathrm{m})^{-1}$. The mean IWV values at all stations on the east coast were up to 5 kg m^{-2} (≈ 20 %) higher than the IWV values in the reference atmosphere at the corresponding heights, while the IWV values on the west coast, particularly at lower station heights, were closer to the values in the reference atmosphere. The magnitude of the diurnal IWV variation at a GPS station was obtained by calculating the mean diurnal IWV evolution averaged over all selected days and by determining the difference between the maximum and minimum IWV values. The variations at the stations on the east coast were mostly stronger than or similar to those on the west coast (Fig. 6.4b). The diurnal evolution of the mean IWV values was compared for Solenzara on the east coast and for Ajaccio

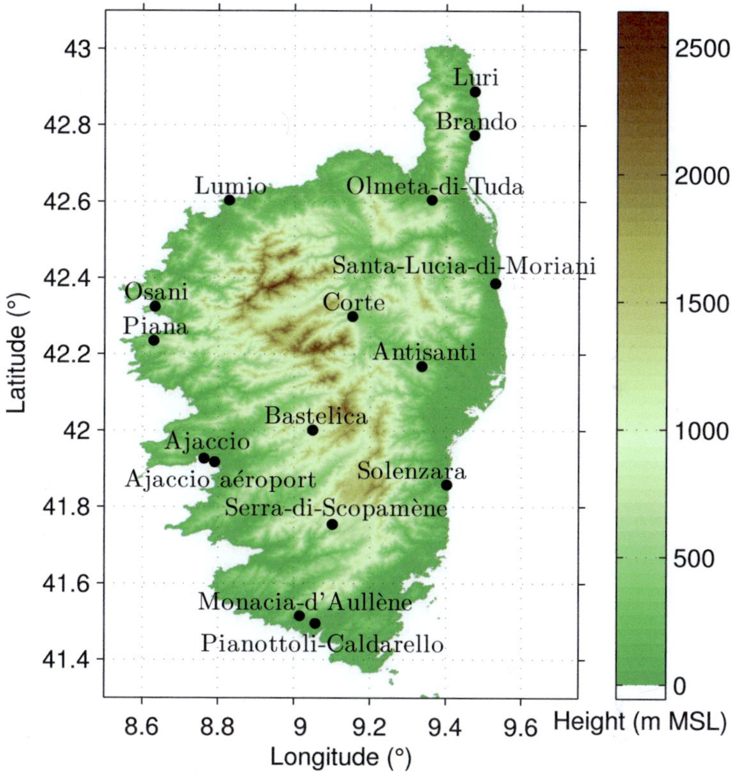

Figure 6.3.: Orography of the Corsican Island and location of operational GPS stations.

on the west coast. The analysis revealed that the atmosphere at Solenzara was generally moister than at Ajaccio, independent of the time of the day (Fig. 6.5a). As the station heights differed only by about 40 m, the humidity offset between the coasts might be related to different coast characteristics, i.e. a rather flat east coast and a steep west coast, or zonal large-scale humidity gradients, which might be evident because of the averaging over a limited number of days. While the difference was mostly less than 3 kg m^{-2} during the morning, the IWV at Solenzara started to increase after 1300 LT, which led to an IWV difference of 5 kg m^{-2} at 1700

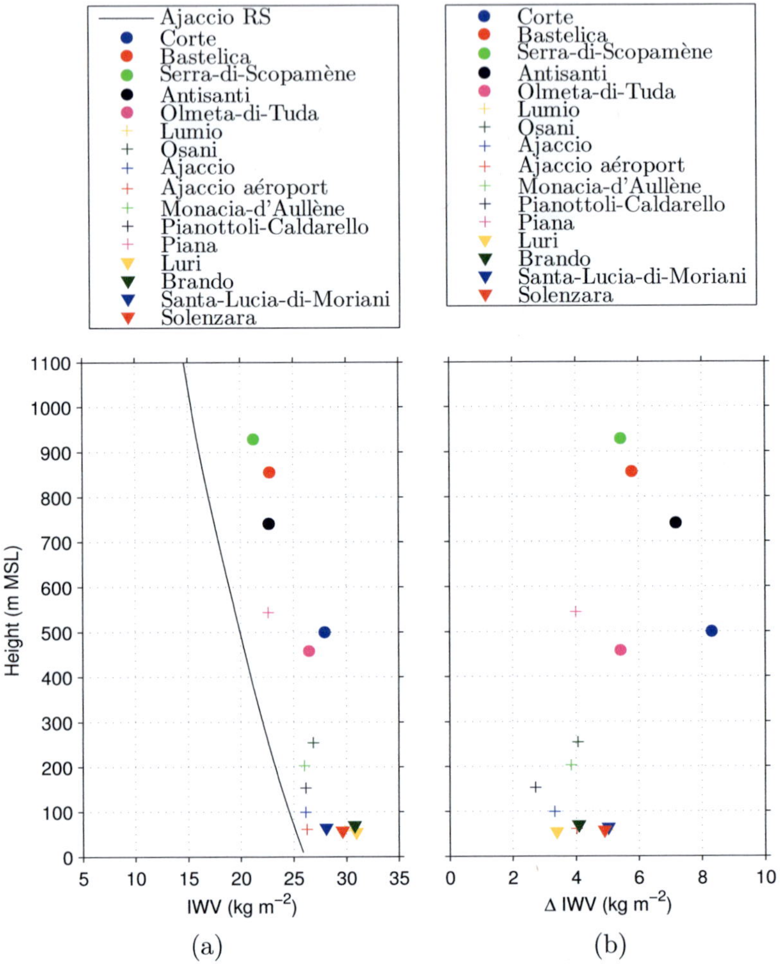

Figure 6.4.: Profiles of IWV in the reference atmosphere measured by radiosondes (RS) at Ajaccio at 1300 LT and height-dependency of IWV derived from GPS measurements averaged between 1300 and 1400 LT (a) and height-dependency of the diurnal IWV variation, Δ IWV, derived from GPS measurements (b). The IWV values are averaged over the days marked by the grey vertical bars in Fig. 6.1. Circles denote stations in the interior of the island, crosses stations on the west coast and triangles stations on the east coast.

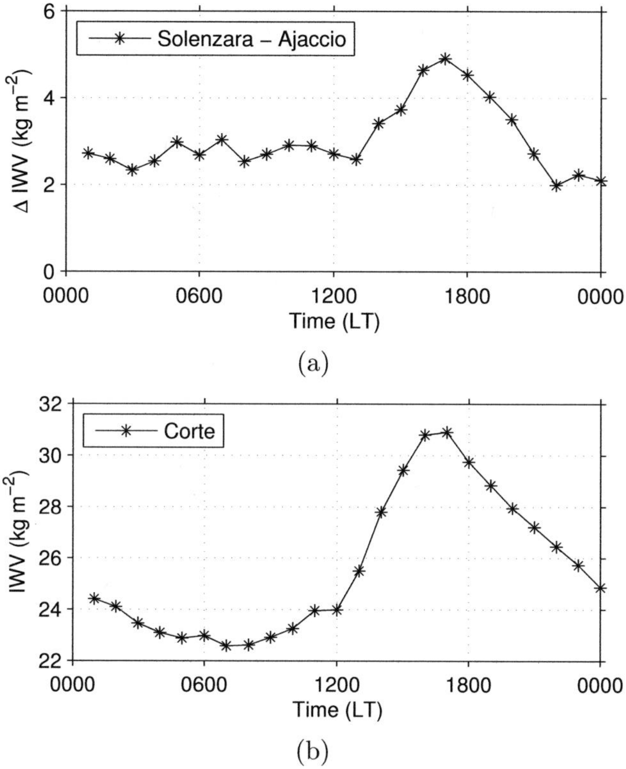

Figure 6.5.: Difference of mean IWV values, Δ IWV, between Solenzara and Ajaccio (a) and mean IWV values at Corte (b). The IWV values are derived from GPS measurements and averaged over the days marked by the grey vertical bars in Fig. 6.1.

LT (Fig. 6.5a). The occurrence of the larger IWV difference in the afternoon suggests that it was related to the existence of a deep mountain ABL and that the east and west coasts were influenced by an unequal mountain ABL evolution.

All stations in the interior of the island measured significantly (> 5 kg m^{-2}) higher IWV values than at the corresponding heights in the reference atmosphere. The stations at Olmeta-di-Tuda and Corte were located at around 450 and 500 m MSL, respectively. Nevertheless, the mean IWV value there was almost as high as at the stations on the coasts at much lower elevations and about 5 kg m^{-2} higher than at Piana, a station at around 550 m MSL on the steep west coast. Because this difference was much higher than the potential contribution due to evapotranspiration, this suggests that even the high GPS stations were influenced by the evolution of a deep mountain ABL. It stands out that the strongest diurnal IWV variation of all stations occurred at Corte (Fig. 6.4b). At the other stations in the interior of the island the IWV variation was weaker than at Corte, but still as high as or higher than at the coast stations. On average, the IWV at Corte started to rise shortly after noon between 1200 and 1300 LT (Fig. 6.5b), i.e. one hour earlier than at Solenzara, reaching maximum values in the late afternoon at around 1600 and 1700 LT. Measurements at Corte also indicated that the diurnal variation of IWV on the selected days was stronger in August than in October, i.e. it depended on the season (Fig. 6.1).

In order to explain the variability of IWV over the Corsican Island, factors responsible for the differences in the water-vapour distribution were analysed by means of radiosonde and microwave radiometer measurements for days with undisturbed mountain ABL evolutions and for a case when isolated deep convection developed above the island. To further illustrate the impact of a mountain ABL on water vapour, the spatial distribution of IWV for days with interrupted and inhibited mountain ABL evolutions was investigated.

6.1.1. Undisturbed Mountain ABL Evolution

The temporal evolution and vertical distribution of water vapour is analysed for days with undisturbed mountain ABL evolutions mostly using the examples of three stations; one in the interior of the island (Corte), one on the east coast (San Giuliano, which is located near the GPS station at Santa-Lucia-di-Moriani) and one on the west coast (Ajaccio). Additionally, the IWV data from the GPS stations at Solenzara and Piana are utilized. To gain information about the vertical distribution of water vapour, IOP days were selected, when radiosondes were launched at Corte and San Giuliano. In order to investigate a potential seasonal dependency of the magnitude of the IWV variation, two days were selected; one in August and one in October. Appropriate IOP days were 24 August and 05 October, when the IWV values resembled the typical characteristics evident in Fig. 6.4.

Case I: 24 August

On 24 August, the IWV at Corte measured by microwave radiometer, radiosondes and GPS showed a strong diurnal variation of more than 12 kg m^{-2} (Fig. 6.6). It decreased during the morning reaching minimum values just before noon, followed by a strong increase after around 1300 LT. Maximum IWV values of more than 30 kg m^{-2} occurred in the late afternoon. After around 2000 LT the IWV decreased again. On the east coast, the IWV measured by microwave radiometer and radiosondes at San Giuliano also showed a diurnal variation with slightly higher maximum values

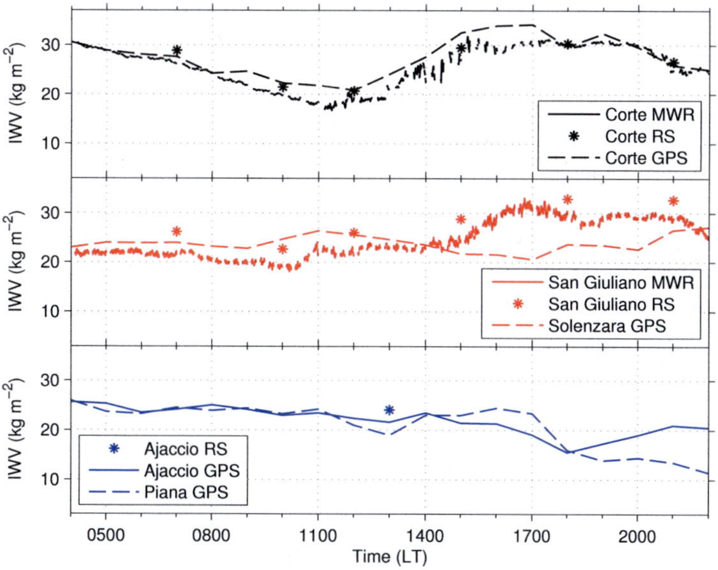

Figure 6.6.: IWV measured by microwave radiometers (MWR) and radiosondes (RS) and derived from GPS measurements at Corte, the east coast and the west coast on 24 August.

than at Corte. Two main increases were found: the IWV rose by several kg m^{-2} between about 1000 and 1100 LT and by about 10 kg m^{-2} between about 1500 and 1700 LT. The IWV values at San Giuliano and further south on the east coast at Solenzara were rather similar before about 1500 LT. Because the IWV at Solenzara did not increase in the course of the afternoon, a north-south difference of about 10 kg m^{-2} developed between the two stations afterwards. On the west coast, no diurnal variations of IWV were detected at the GPS stations neither at Ajaccio nor at Piana. In the late afternoon and evening, both stations detected an IWV decrease. Consequently, the maximum IWV values on the west coast remained about 10 kg m^{-2} lower than at Corte and San Giuliano. Because the radiosonde at Ajaccio only reached about 3800 m MSL, the water-vapour density was linearly extrapolated above, assuming a water-vapour density of 0 g m^{-3} at 12000 m MSL. Next, the radiosonde profiles were investigated to determine the layers which were responsible for the diurnal cycle of IWV at Corte and San Giuliano.

Corte In the early morning, a moist layer of several hundred metres of depth existed in the valley centre (Fig. 6.7a). The subsequent IWV decrease mainly resulted from a humidity decrease below about 900 m MSL. Starting at around 0700 LT, a scanning microwave radiometer detected horizontal differences in the IWV distribution with higher IWV values above the ridges south-west of Corte and lower values in the valley (Fig. 6.7b). This suggests that the decrease of IWV and humidity close to the surface in the valley centre at Corte in the morning was caused by topographic

Figure 6.7. ***(facing page)*****:** Water-vapour density (colour-coded), potential temperature (black isolines) and horizontal wind vector (arrows) measured by radiosondes at Corte (a) and San Giuliano (c) and azimuth time plot of IWV measured by a microwave radiometer at Corte (b) on 24 August. In (a) and (c), water-vapour density is interpolated using microwave radiometer and radiosonde data, potential temperature is temporally linearly interpolated between the radiosoundings and the mountain ABL height is indicated by the thick dashed line. In (b), the vertical black lines mark the azimuth angle sector referred to in the text.

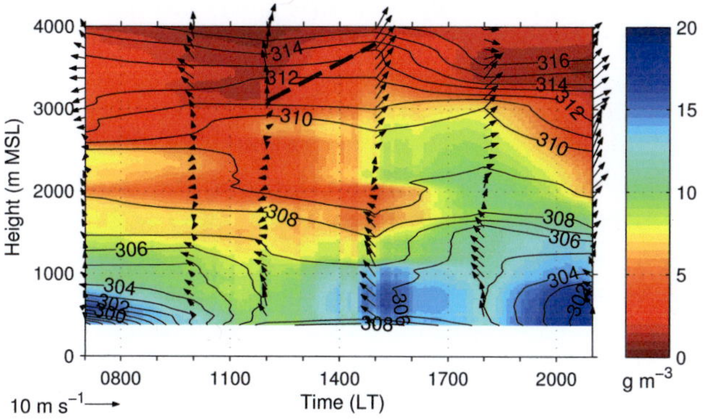

Height (m MSL)
Time (LT)
10 m s⁻¹
g m⁻³

(a)

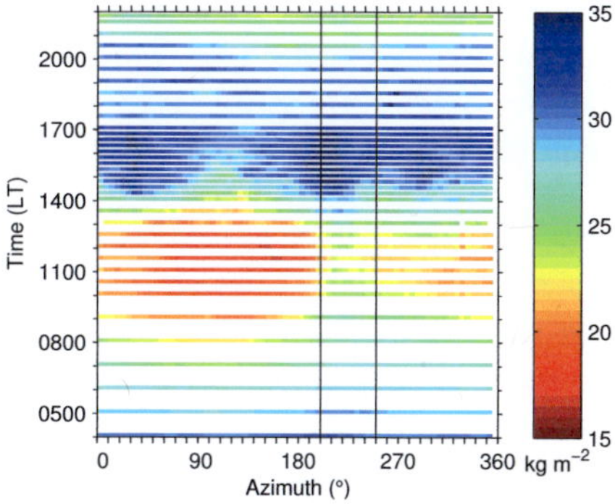

Time (LT)
Azimuth (°)
kg m⁻²

(b)

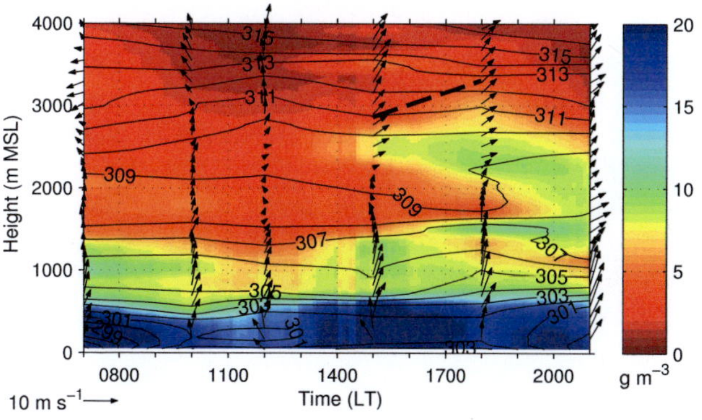

Height (m MSL)
Time (LT)
10 m s⁻¹
g m⁻³

(c)

venting, i.e. the transport of water vapour close to the surface with upslope winds from the valley centre up the slopes over the ridges.

After noon, significant accumulations of water vapour could be distinguished in two different layers (Fig. 6.7a). The water-vapour density increased by more than 5 g m^{-3} below about 800 m MSL after 1300 LT, which could rise the IWV at the surface by 4 kg m^{-2}. The moistening probably resulted from the arrival of the sea-breeze front at Corte. In addition, the water-vapour density started to increase mostly below about 3000 m MSL after around 1200 LT. The rise was first detected in an elevated shallow layer and successively extended to lower layers during the next couple of hours and gained more than 5 g m^{-3}. As the evolution of a mountain ABL was determined via the elevated humidity increase (Sect. 4.2), a mountain ABL up to about 3800 m MSL at 1500 LT was detected at Corte. The top of the mountain ABL was accompanied by a stronger stable stratification above. The horizontal differences in the humidity distribution with higher humidity values above the ridges surrounding Corte intensified during the morning (Fig. 6.7b). At 1200 LT, when the water-vapour density increase started in the elevated layer above Corte, the IWV was up to 5 kg m^{-2} higher in the south-west (200° to 250°) of Corte compared to other directions and compared to the IWV above Corte (Fig. 6.6). As the layer with the humidity increase above Corte was accompanied by weak south-westerly to westerly wind (Fig. 6.7a), advective venting was likely to occur and to cause the strong moistening and deep mountain ABL at Corte in the afternoon.

Before a deep mountain ABL was detected at Corte at 1200 LT (Fig. 6.7a), the profiles of IWV at Corte and Ajaccio were very similar (Fig. 6.8a). Afterwards, the IWV profiles at Corte differed significantly from the one at 1200 LT with higher values in the mountain ABL (Fig. 6.8a). Assuming that the IWV profile at Ajaccio at 1300 LT was representative for the whole afternoon, horizontal humidity differences in a layer of more than 3000 m depth developed between Ajaccio and Corte.

San Giuliano The IWV variations at San Giuliano also resulted from water-vapour density changes in different layers (Figs. 6.6 and 6.7c). Affected by a south-easterly sea breeze, the water-vapour

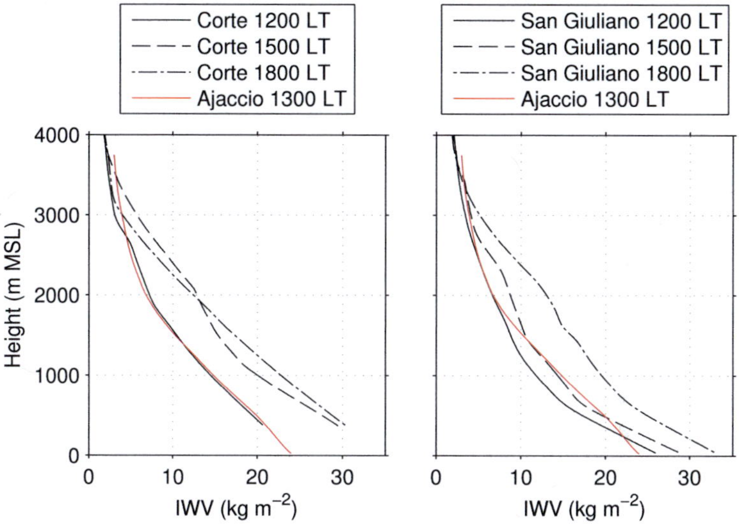

(a) (b)

Figure 6.8.: Profiles of IWV measured by radiosondes at Corte (a) and San Giuliano (b) on 24 August. In each plot, the IWV profile measured by radiosonde at Ajaccio at 1300 LT on 24 August is shown.

density increased by about 3 g m^{-3} in a moist layer below about 500 m MSL between 1000 and 1500 LT, which contributed to an IWV rise of 1.5 kg m^{-2}. After around 1400 LT, the humidity started to increase in an elevated layer at around 2500 m MSL and successively expanded to lower layers (Fig. 6.7c). The water-vapour density increased by about 5 g m^{-3} in a layer of about 2000 m depth until 2100 LT, compared to the conditions at noon. This contributed to an IWV increase of 10 kg m^{-2}. With south-westerly to westerly flow in the layer where the water-vapour density increased, San Giuliano was on the downstream site of the mountain ridges and therefore could have been affected by advective venting of moist air from the ridges. This assumption allowed the determination of a deep mountain ABL at San Giuliano, which extended to about 3300 m MSL at 1800 LT.

Due to the late evolution of a deep mountain ABL at San Giuliano, the IWV profile at 1500 LT differed only little from the

IWV profile at noon and from the IWV profile at Ajaccio at 1300 LT (Fig. 6.8b). Once a deep mountain ABL had evolved and the water-vapour density had increased at San Giuliano, the IWV at 1800 LT revealed a similar profile as at Corte in mid- and late afternoon (Fig. 6.8a).

Ajaccio At Ajaccio, a westerly wind component below about 800 m MSL was detected at 1300 LT, which was probably associated with a sea breeze. The water-vapour density in this layer was significantly lower than at San Giuliano (Fig. 6.8b), which might have been related to different coast characteristics, as the west coast was much steeper than the east coast. Above about 800 m MSL, an easterly wind component existed up to about 3500 m MSL, which could have allowed the advection of moist air from the mountain ridges of the Corsican Island towards the west coast, although the IWV did not show an increase during the day (Fig. 6.6). Unfortunately, one available radiosonde during daytime was not sufficient to examine this hypothesis.

Case II: 05 October

On 05 October, the diurnal cycles of IWV at Corte and San Giuliano were much less pronounced (Fig. 6.9) compared to 24 August (Fig. 6.6). At Corte, the IWV increased by about 5 kg m^{-2} between about 1400 and 1600 LT and remained high until about 1800 LT, while the IWV variations did not exceed 2 kg m^{-2} at any station on the east coast. Again, no strong diurnal variations occurred on the west coast. Because of a large-scale west-east humidity gradient, the IWV values on the west coast were about 5 kg m^{-2} lower than on the east coast. The absolute IWV values were in general lower than on 24 August and did not exceed 20 kg m^{-2} at any of the stations.

Corte Vertical profiles revealed that the water-vapour density at Corte increased within the combined upvalley-wind and sea-breeze layer below about 1200 m MSL by more than 3 g m^{-3} after about 1400 LT (Fig. 6.10a). Approximately at the same time, the water-vapour density increased above that layer up to about 2600 m MSL. Higher IWV in a sector between 170° and 250°, i.e. roughly above the ridges south-west of Corte, compared to the surroundings was

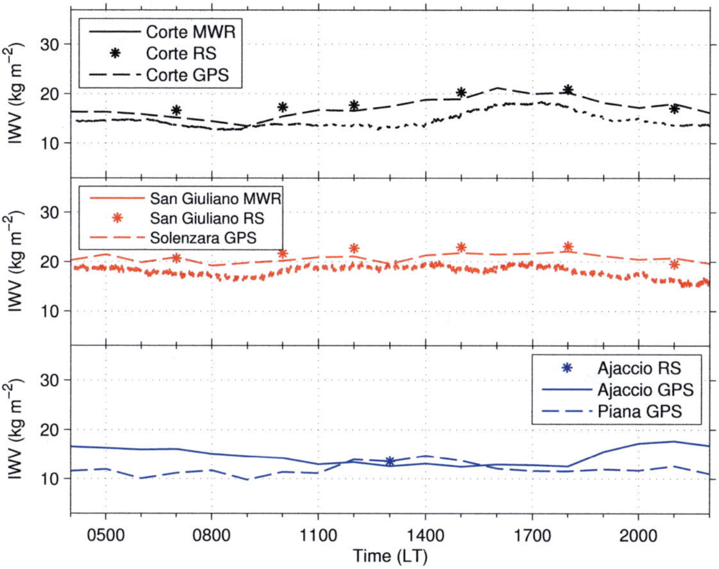

Figure 6.9.: IWV measured by microwave radiometers (MWR) and radiosondes (RS) and derived from GPS measurements at Corte, the east coast and the west coast on 05 October.

measured by the scanning microwave radiometer after around 1100 LT (Fig. 6.10b). The spatial differences became more pronounced in the course of the afternoon. As the wind direction was southerly above the combined upvalley-wind and sea-breeze layer turning to westerly with height (Fig. 6.10a), advective venting of moist air masses from the ridges south-west of Corte was likely and the mountain ABL deepened significantly. At 1800 LT, the water-vapour density was up to 4 g m^{-3} higher in a layer between about 1200 and 2200 m MSL compared to the sounding at noon. The moistening in both layers contributed to an IWV increase of more than 6 kg m^{-2} until 1800 LT. However, large-scale advection of dry air above about 3000 m MSL lowered the IWV increase.

The large-scale humidity gradient was reponsible for the offset of the IWV profiles at Corte and Ajaccio at 1200 and 1500 LT (Fig.

6.11a). Until 1800 LT, large-scale advection of dry air removed this gradient above about 2200 m MSL, while the evolution of a deep mountain ABL at Corte sustained a horizontal humidity gradient below that height.

San Giuliano A weak increase in water-vapour density of about 1 g m^{-3} in the lower 800 m MSL occurred at San Giuliano at around 1100 LT, which was likely associated with the onset of the south-easterly sea breeze (Fig. 6.10c). However, this moisture supply contributes less than 1 kg m^{-2} to the IWV evolution. Apart from that, almost no humidity increase occurred at San Giuliano during the day. Only in the evening, water-vapour density increased by up to 2 g m^{-3} in an elevated layer between about 1600 and 2400 m MSL. As a westerly wind component existed in this layer, advective venting of moist air masses from the upstream mountain ridges could have occurred. Below this layer, the wind direction was mainly southerly, thus ridge parallel, which presumably prevented advective venting of moist air from the mountains. Because of the simultaneous occurrence of large-scale dry air advection above about 3000 m MSL (Fig. 6.10c), the humidity increase in the elevated layer did not show in the IWV evolution (Figs. 6.9 and 6.11b).

Ajaccio Similar to 24 August, a westerly wind component below about 700 m MSL and an easterly wind component between about 700 and 1300 m MSL existed at Ajaccio at 1300 LT. The latter could have allowed advective venting of moist air from the mountain

---

Figure 6.10. ***(facing page)***: Water-vapour density (colour-coded), potential temperature (black isolines) and horizontal wind vector (arrows) measured by radiosondes at Corte (a) and San Giuliano (c) and azimuth-time plot of IWV measured by a microwave radiometer at Corte (b) on 05 October. In (a) and (c), water-vapour density is interpolated using microwave radiometer and radiosonde data, potential temperature is temporally linearly interpolated between the radiosoundings and the mountain ABL height is indicated by the thick dashed line. In (b), the vertical black lines mark the azimuth angle sector referred to in the text.

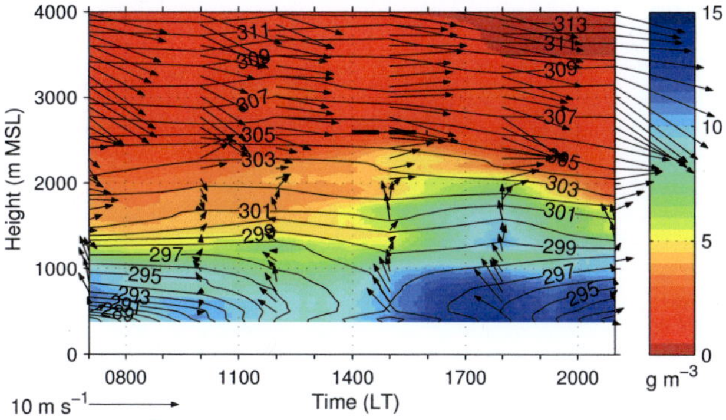
Height (m MSL)
4000
3000
2000
1000
0
0800
1100
1400
1700
2000
Time (LT)
10 m s⁻¹
g m⁻³
15
10
5
0

(a)

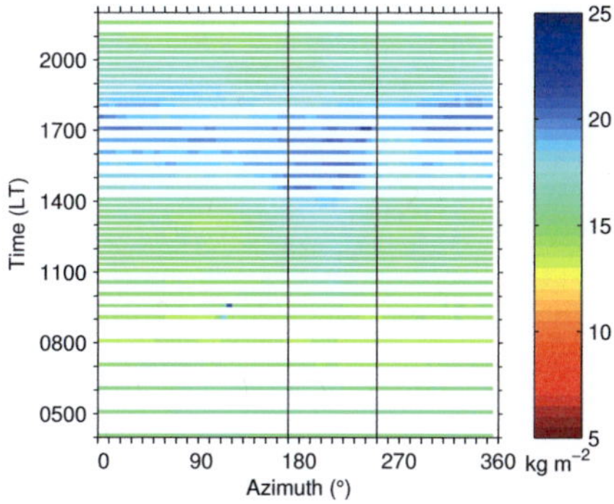
Time (LT)
2000
1700
1400
1100
0800
0500
0
90
180
270
360
Azimuth (°)
kg m⁻²
25
20
15
10
5

(b)

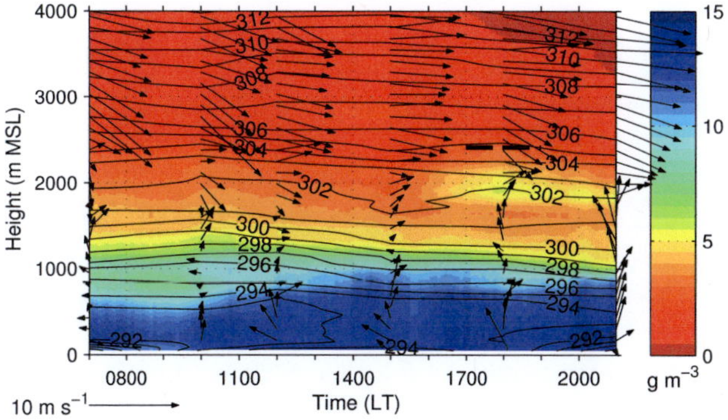
Height (m MSL)
4000
3000
2000
1000
0
0800
1100
1400
1700
2000
Time (LT)
10 m s⁻¹
g m⁻³
15
10
5
0

(c)

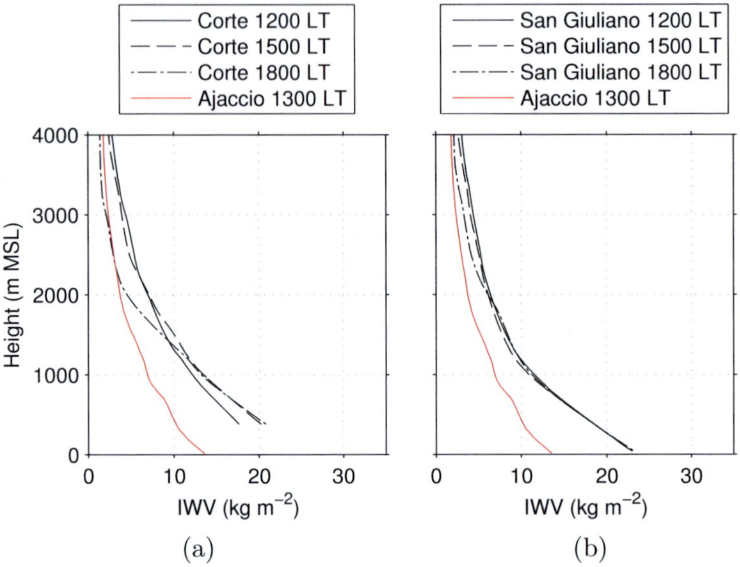

Figure 6.11.: Profiles of IWV measured by radiosondes at Corte (a) and San Giuliano (b) on 05 October. In each plot, the IWV profile measured by radiosonde at Ajaccio at 1300 LT on 05 October is shown.

ridges towards the coast, although there was no evidence for that in the IWV evolution (Fig. 6.9).

Seasonal Dependency of the Water-Vapour Distribution

At Corte, the diurnal IWV increase started earlier in August (1300 LT) than in October (1400 LT) and the diurnal IWV variation was larger in August ($\approx$ 12 kg m^{-2}) than in October ($\approx$ 5 kg m^{-2}). The small diurnal IWV variation on 05 October was at least partly caused by large-scale advection of dry air (Figs. 6.6 and 6.9). Because thermally driven circulations associated with the evolution of a deep mountain ABL were identified to be mainly responsible for the diurnal variation of IWV, the seasonal dependency of these circulations was investigated. In agreement with a higher solar radiation in August, the surface sensible heat flux at Corte was almost 200 W m^{-2} higher than in October (Fig. 6.12). The

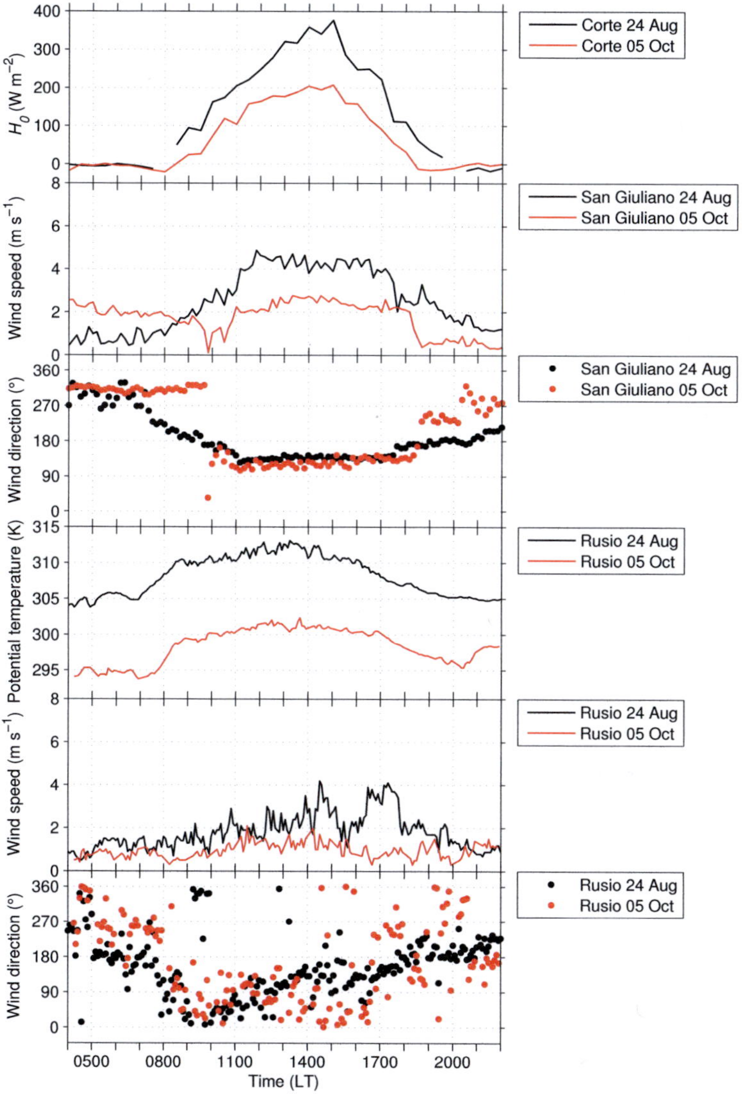

Figure 6.12.: Surface sensible heat flux, H_0, at Corte; near-surface wind speed and wind direction at San Giuliano; and temperature and wind speed and direction at Rusio on 24 August and 05 October.

strength of the surface sensible heat flux affects the near-surface temperature and the intensity and duration of thermally driven circulations. The onset of the sea breeze at the coast was somewhat earlier and the wind speed in the sea-breeze layer was stronger in August (detected at San Giuliano, Fig. 6.12). Thus, the sea-breeze front could have propagated faster along the axis of the Tavignano Valley and arrived earlier at Corte, causing the earlier humidity increase in the combined upvalley-wind and sea-breeze layer (Figs. 6.7a and 6.10a). Upslope winds existed in August as well as in October, as observed at Rusio (Fig. 6.12) and in azimuth and elevation scans performed by the wind lidar at Corte. However, they were stronger and their onset was about one hour earlier in August.

In the analysed case studies, the mountain ABL in August was about 1200 m deeper than in October. Parameters affecting the top of the mountain ABL are assumed to be mountain-ridge height, near-surface temperature and atmospheric stratification (Sect. 5.1). The near-surface potential temperature at Rusio reached maximum values of about 312 K on 24 August and about 302 K on 05 October (Fig. 6.12). Together with a weak stable stratification above the mountain ridges (Fig. 6.7a), the higher near-surface potential temperature on 24 August allowed an air parcel to rise dry-adiabatically up to about 3400 m MSL, before it became negatively buoyant. Because of the lower near-surface temperature and strong stable stratification on 05 October (Fig. 6.10a), an air parcel would reach its level of neutral buoyancy at about 2200 m MSL, which was 1200 m lower than on 24 August. The estimated levels of neutral buoyancy were somewhat lower than the detected mountain ABL heights for both days. Nevertheless, the same seasonal height difference was evident for both heights and proves the assumed dependencies listed above.

Because the water-vapour density in the upvalley-wind layer, associated with the sea-breeze front arrival, increased earlier and the upslope winds were stronger in August, topographic and advective venting of moist air could also occur earlier in August. The stronger upslope winds only have a positive effect on the IWV increase, as long as the humidity gradient between the upvalley-wind layer and the elevated layers is maintained. Once it vanishes or weakens, advection cannot or can only little increase the water-vapour density in the elevated layers and therefore the IWV. However, the major parameter controlling the diurnal IWV variation was presumably the mountain ABL depth. Assuming an equal stable stratification, a higher near-surface potential temperature generally leads to a deeper mountain ABL. Thus, the water-vapour density increase in August could extent over a deeper layer and could cause a stronger IWV increase.

6.1.2. Isolated Deep Convection

Using the example of 23 August, the water-vapour distribution was investigated for a day when isolated deep convection occurred over the Corsican Island. Until mid-afternoon, the IWV at Corte revealed the typical diurnal evolution (Fig. 6.13) which was associated with a deep mountain ABL, similar to the one observed

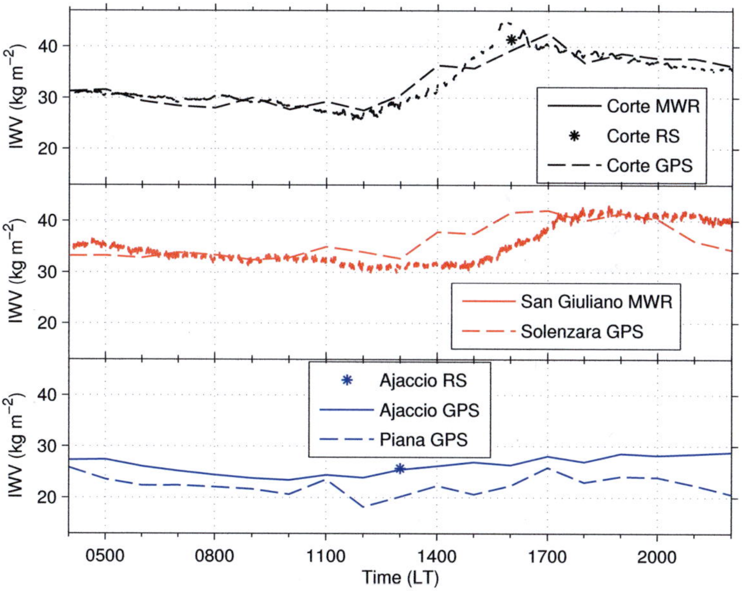

Figure 6.13.: IWV measured by microwave radiometers (MWR), radiosondes (RS) and GPS at Corte, the east coast and the west coast on 23 August.

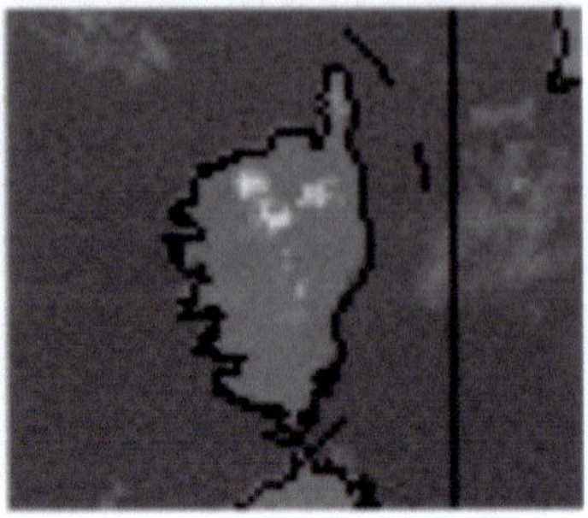

(a)

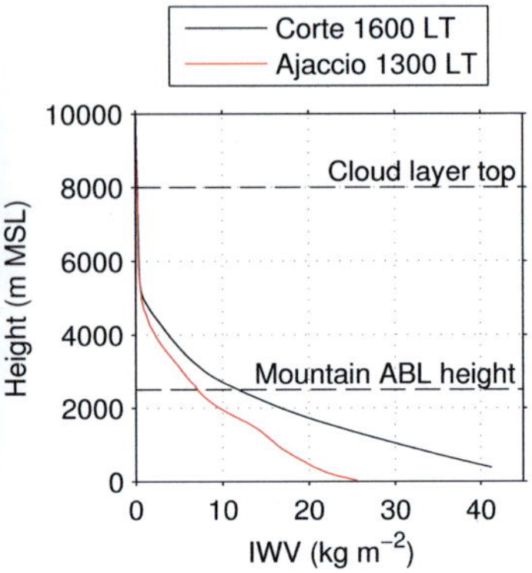

(b)

Figure 6.14.: MSG satellite image (courtesy of EUMETSAT) at 1500 LT (a) and profiles of IWV measured by radiosondes at Corte and Ajaccio (b) on 23 August. In (b), the mountain ABL height, estimated using the near-surface potential temperature at Rusio and the potential temperature profile at Ajaccio, and the cloud-layer top measured by cloud radar at Corte are indicated by the dashed horizontal lines.

on 24 August. The height of the mountain ABL was estimated to about 2500 m MSL, using the near-surface potential temperature at Rusio and the potential temperature profile measured by a radiosonde launched at Ajaccio at 1300 LT, as no radiosonde at Corte was available at this time.

After about 1430 LT, three isolated deep moist convective cells developed above the mountains in the northern part of the Corsican Island as detected by a rain radar positioned at San Giuliano and visible in a MSG satellite image (Fig. 6.14a). One of the cells was initiated above the mountains south-west of Corte and got advected above the measurement station at Corte, as visible in cloud-camera images. When it passed the station between about 1530 and 1600 LT, the cloud radar measured a cloud-layer top higher than 8000 m MSL and a cloud-base height between about 2500 and 3000 m MSL was derived from the cloud-base temperature measured by a infrared radiometer and the temperature profile measured by radiosonde at Corte at 1600 LT. The passage of the cell was accompanied by an IWV increase of about 4 kg m^{-2} (Fig. 6.13).

Possible processes to cause this increase are horizontal advection of moister air or phase changes from solid or liquid into vapour within the cloud. As the cloud-camera images indicated that the deep convective cell was already dissolving when it passed Corte, the latter process was probably significant.

Right after the passage of the deep convective cell a radiosonde was launched at Corte. Comparing the IWV profiles from soundings at Corte and at Ajaccio shows that the atmosphere at Corte below about 5000 m MSL was rather moist, leading to higher IWV by about 20 kg m^{-2} at about 370 m MSL, i.e. the height of the station at Corte (Fig. 6.14b). As the mountain ABL reached only to about 2500 m MSL and no large-scale humidity differences were evident, the moist layer above that height could be attributed to vertical transport processes in the deep convective cell itself.

6.1.3. Interrupted and Inhibited Mountain ABL Evolution

The relation between the IWV values derived from GPS measurements and the IWV profile in the reference atmosphere on 02 and 09 October strongly differed from the typical relation evident in Fig. 6.4a. Between 1600 and 1700 LT on 02 October, dry air was found to intrude the mountain ABL at Corte (Sect. 5.2). In this time period, the IWV derived from GPS measurements at Corte was very similar to the one in the reference atmosphere (Fig. 6.15a), indicating that no deep mountain ABL affected the water-vapour distribution above the site. At the other sites, an impact of a deep mountain ABL was still present, as the IWV relation between most GPS stations and the reference atmosphere approximately resembled the typical relation (Fig. 6.4a).

As outlined in Sect. 5.3, very strong westerly wind prevailed and dry air from the free atmosphere intruded into the atmosphere downstream of the main mountain ridge during the morning and early afternoon of 09 October. This inhibited the evolution of a mountain ABL at Corte. A moist shallow layer on the upstream side of the island was blocked from overpassing. These conditions were well reflected by the IWV distribution (Fig. 6.15b): due to the blocking the west coast was moist, resulting in higher IWV values than at most east coast sites. The IWV values at the stations in the interior of the island were mainly close to those in the reference

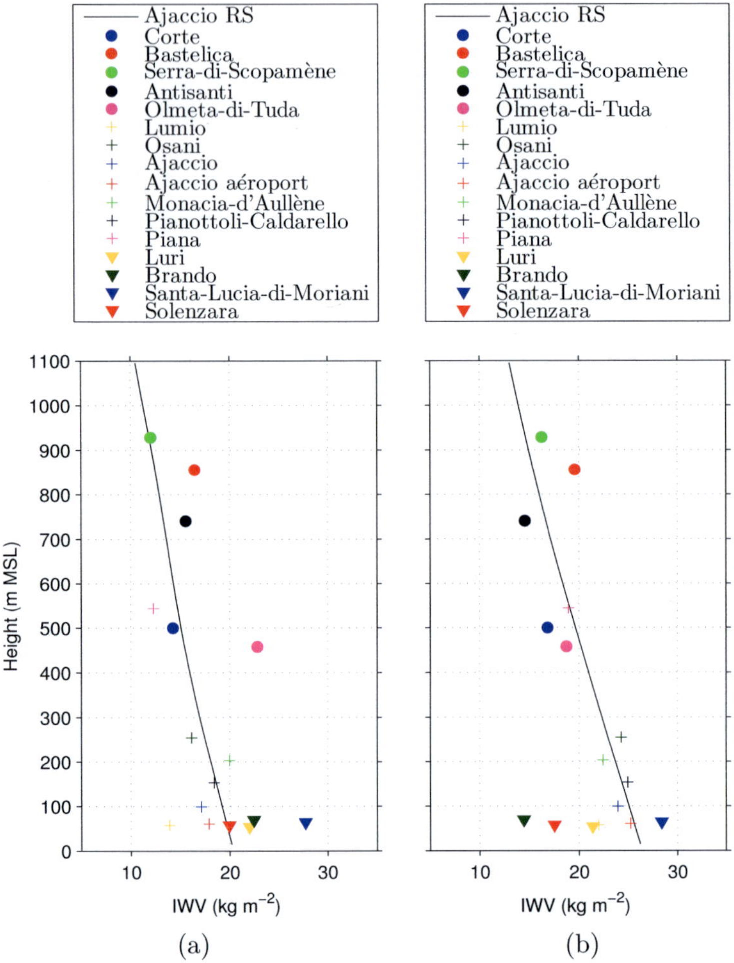

Figure 6.15.: Profiles of IWV measured by radiosondes (RS) at Ajaccio at 1300 LT and height-dependency of IWV derived from GPS measurements averaged between 1600 and 1700 LT on 02 October (a) and 1300 and 1400 LT on 09 October (b). Circles denote stations in the interior of the island, crosses stations on the west coast and triangles stations on the east coast.

atmosphere, which reveals that no deep mountain ABL was able to develop over a large part of the island. The rather high IWV at Bastelica compared to the reference atmosphere was probably owed to the station location west of the main mountain ridge (Fig. 6.3) and upward transported moist air from the blocked layer along the upstream mountain slopes.

6.2. Convection-Related Parameters

The statistical analysis of convection-related parameters and the occurrence of isolated deep convection over the island indicated that the conditions on the west coast were not representative for the atmospheric conditions, in which deep convection evolved (Sect. 1). To identify days with deep convection the cloud-to-ground lightning activity was used as an indicator for deep convection. Figure 6.16 shows the distribution of the lightning activity averaged over the days considered in the statistical analysis. Strong spatial inhomogeneities are evident: a maximum over and to the east of the Corsican Island occurred. As westerly flow conditions dominated over the Corsican Island as evident from Fig. 6.2 and also found by Lambert et al. (2011), the lightning intensity was concentrated above and downstream of the main mountain ridge. Assuming that the lightning activity indicates deep convection and precipitation events, this was likely caused by deep convective cells, which were triggered over the mountains and downstream advected with the mean wind. This also means that preferred regions for the occurrence of deep convection are over and downstream of the mountains - in areas generally affected by a deep mountain ABL.

In order to analyse the impact of the mountain ABL on the pre-convective atmospheric conditions, the temporal evolution of convection-related parameters on days with different mountain ABL evolutions was investigated at Corte in the interior of the island. Spatial differences of convection-related parameters on days with and without deep convection were analysed using data from this station, which was located very close to the location where the maximum of deep convection occurred, and from Ajaccio on the west coast, which is found in some distance from most deep convective cells (Fig. 6.16). Furthermore, the relation between

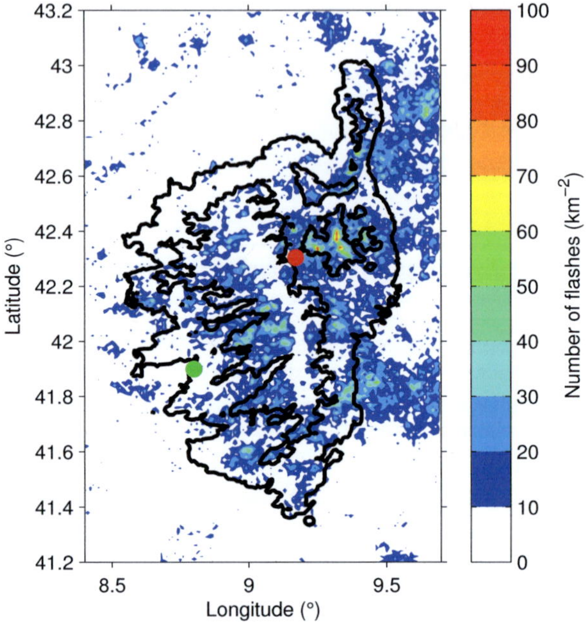

Figure 6.16.: Number of cloud-to-ground flashes between 1200 and 2400 LT for the months August, September and October from 2001 to 2010. The data derive from the Siemens lightning information service (BLIDS), which is based on the EUropeean Cooperation for LIghtning Detection (EUCLID). The orography of Corsica is indicated by the black isolines and the locations of Ajaccio and Corte are marked by the green and red dots, respectively.

the conditions at the different sites and the evolution of deep convection was assessed. Days without deep convection include IOP days with undisturbed mountain ABL evolutions (15 August; 19 August, Sect. 5.1; 24 August; 21 September; 23 September; and 05 October) as well as days when the mountain ABL evolution was interrupted (02 October, Sect. 5.2) or inhibited (09 October, Sect. 5.3) by the background flow (see also Tab. B.1 in the appendix). Unfortunately, of all days with isolated deep convection (23 August, 25 August, 06 September and 09 September) only 06 September was an IOP day with regular radiosoundings at Corte. Available

Stüve diagrams for the analysed days at around noon are included in the appendix (Figs. B.6 to B.17).

6.2.1. Temporal Evolution at Corte

CAPE and CIN

As CAPE and CIN could only be obtained when temperature and humidity profiles were available from radiosoundings, these parameters were compared for IOP days. A diurnal evolution of CAPE and CIN at Corte only occurred on three days with undisturbed mountain ABL evolutions, i.e. on 19 August, 24 August and 23 September. On these days, CAPE increased during the day and reached maximum values in the afternoon (Fig. 6.17). Simultaneously, CIN decreased. On 15 August, a lid at around 3500 m MSL prevented high CAPE values in the afternoon, while on 06 September, i.e. a day with deep convection, CAPE increased and CIN even became zero in the course of the day. On the other IOP days, CAPE and CIN either showed no diurnal variation or could not be calculated for different reasons: on 21 September and 05 October a lifted air parcel never got positively buoyant and thus CAPE and CIN were not defined. On 02 and 09 October

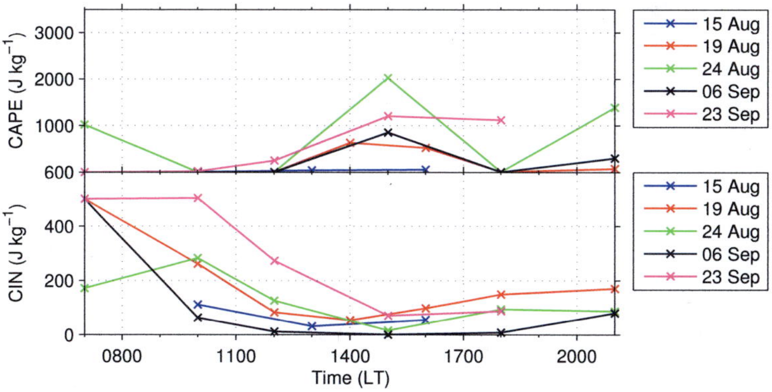

Figure 6.17.: CAPE and CIN derived from radiosonde profiles at Corte on 15 August, 19 August, 24 August, 06 September and 23 September.

no diurnal variation of CAPE and CIN could be detected, as the conditional instability was either limited due to elevated inversions or not defined.

TT Index and RH

Furthermore, the TT index and RH were used to describe atmospheric stratification and low- and mid-level humidity. These parameters had the advantage of existing for each profile and could also be calculated from microwave radiometer profiles for days when no radiosonde profiles were available. Because RH was averaged between about 1500 and 3000 m MSL, i.e. around and

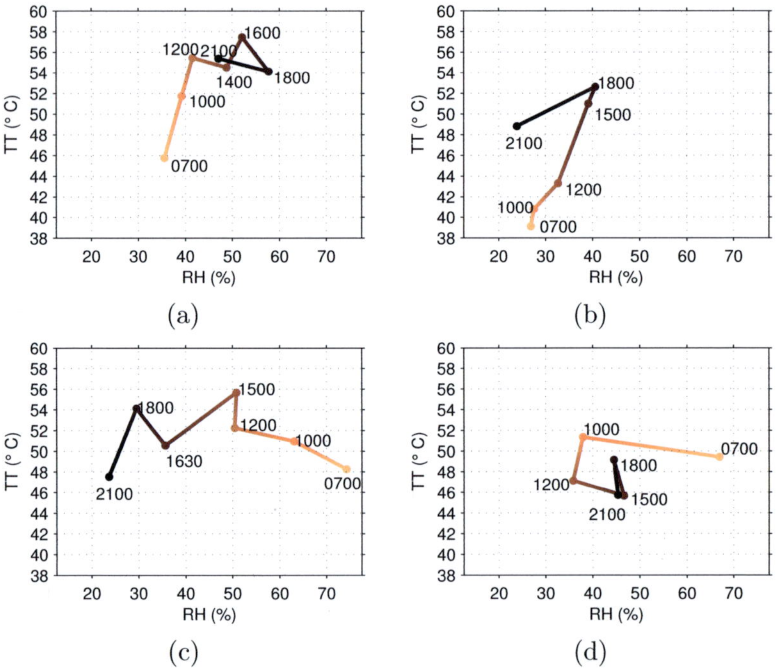

Figure 6.18.: TT index and RH at Corte on 19 August (a), 05 October (b), 02 October (c) and 09 October (d). The parameters are derived from profiles measured by radiosondes and the numbers in the diagrams denote time in LT.

above mountain-ridge height, it reflected the evolution of the upper part of the mountain ABL. When the mountain ABL height was significantly lower than 3000 m MSL, RH could also be dominated by other processes, like large-scale advection. A characteristic evolution of the TT index and RH was evident for days without deep convection and with an undisturbed mountain ABL evolution. In Figs. 6.18a and b, this characteristic evolution is exemplified for 19 August and 05 October. The TT index mostly increased during the day reaching maximum values of 57 °C in August (Fig. 6.18a) and 53 °C in October (Fig. 6.18b) in mid-afternoon or early evening, mainly because of a warming and moistening of the lower troposphere. Starting in mid-morning RH increased, with maximum values of 58 % in August (Fig. 6.18a) and 41 % in October (Fig. 6.18b) occurring in the late afternoon or early evening. This was mainly related to an increase of humidity in the middle troposphere due to the evolution of a deep mountain ABL. This means that the most favourable pre-convective conditions, i.e. high TT index and RH, occurred at about the same time. The evolution of RH was similar to the one of IWV, which confirms the finding that the IWV evolution was significantly influenced by advective venting on days with undisturbed mountain ABL evolutions (Sect. 6.1). Similar to the seasonal evolution of the water-vapour distribution, the TT index and RH depended on the time of the year, with in general higher TT indices and RH values in August than in October (Figs. 6.18a and b).

On 02 October, RH decreased during the course of the day mainly due to large-scale advection of dry air (Fig. 6.18c and Sect. 5.2). Between 0700 and 1200 LT, it dropped by more than 25 %. With the evolution of a deep mountain ABL between 1200 and 1500 LT, the decrease of RH was paused and the TT index strongly increased by almost 4 °C up to about 56 °C. When the evolution of the mountain ABL was interrupted and dry and warm air intruded the mountain ABL around 1630 LT, the decrease of RH was enhanced and the TT index decreased by about 5 °C.

On 09 October, RH was low and the TT index did not increase as long as the intrusion of dry and warm air from the free troposphere inhibited the evolution of a mountain ABL at Corte (Fig. 6.18d and Sect. 5.3). As soon as the intrusion weakened in the afternoon, RH increased until 1500 LT, mostly because the temperature decreased in the relevant layer. Once the moist upvalley-wind layer replaced the intruding dry air after around 1630 LT, the TT index increased.

Overall, the pre-convective environment at Corte considerably depended on the evolution of a mountain ABL, which in general favoured the subsequent development of moist convection, i.e. it decreased the stability of the atmospheric stratification and increased the low- and mid-level humidity content. An intrusion of warm and dry air generally decreased RH and the TT index, which was unfavourable for the evolution of moist convection.

6.2.2. Spatial Distribution over the Corsican Island

Like for Corte, CAPE and CIN were rather often not defined for radiosonde profiles obtained at Ajaccio. Because of that, the TT index and RH were applied to investigate spatial differences in the atmospheric conditions between Corte and Ajaccio. Figure 6.19 shows the TT index and RH on days with an undisturbed mountain ABL evolution, calculated from the 1300 LT radiosounding at Ajaccio and from the 1300, 1400 or 1500 LT radiosounding at Corte. The TT index and RH were always higher at Corte than on the west coast, which was probably caused by the negative impact of the sea breeze on the coast leading to a cooler lower troposphere and by the higher humidity in low- and mid-levels at Corte. Thus, it is concluded that the conditions in the early afternoon were in general more favourable at Corte than on the west coast, which was obviously strongly related to the positive impact of the evolution of a deep mountain ABL at Corte and the negative impact of the cool sea-breeze layer on the coast.

The TT index and RH at Corte and Ajaccio were compared for days with and without deep convection. In addition to radiosonde profiles, the parameters were calculated from microwave radiometer profiles at Corte at 1300, 1400 and 1500 LT. The TT index at both stations varied mainly between about 40 and 60 °C (Fig. 6.20). While days without deep convection were usually associated with a TT index at Ajaccio lower than about 53 °C, deep convection occurred on days with a TT index ranging from about 51 to 61 °C. The RH values at Ajaccio mostly varied between about 10 and 50 % and were rather independent of the occurrence of deep convection, i.e. days with and without deep convection were

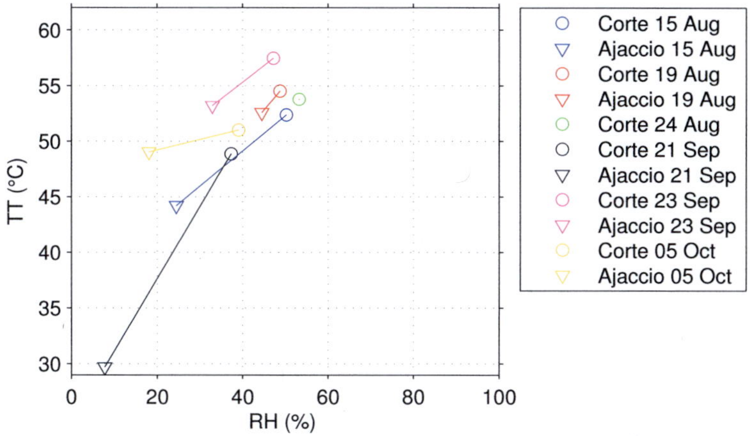

Figure 6.19.: TT index and RH at Corte and Ajaccio on days with undisturbed mountain ABL evolutions. The parameters are derived from radiosoundings at Ajaccio at 1300 LT and at 1300, 1400 or 1500 LT at Corte.

associated with the same RH values. In the statistical analysis in Sect. 1, days with deep convection were classified in four categories using thresholds for the TT index and RH to distinguish between conditions favourable and unfavourable for the evolution of deep convection. During the HyMeX field campaign, four days with isolated deep convection occurred, each of them assignable to another category (Fig. 6.20). This means that deep convection occurred rather independent of the convection-related parameters calculated for Ajaccio. For example on the day of category IV (25 August), deep convection developed above the island despite a TT index of about 51 °C and a RH value of about 25 % at Ajaccio around noon, i.e. the values were much lower than the respective thresholds and did not at all indicate conditions favourable for deep convection.

At Corte, RH varied between about 25 and 90 %, while deep convection only occurred when the RH value was larger than 45 %. In addition, days with deep convection were mostly associated with a TT index higher than the threshold of 57 °C. For RH values ranging between about 45 and 50 % and TT index values between

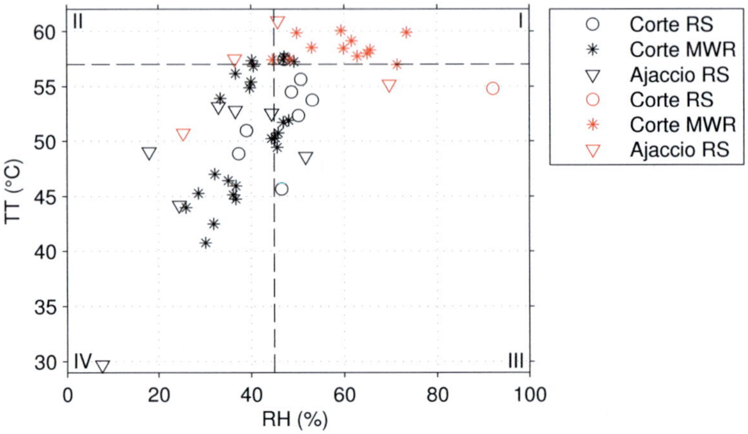

Figure 6.20.: TT index and RH at Corte and Ajaccio on days with (red) and without (black) isolated deep convection. The parameters are derived from profiles measured by radiosondes (RS) at 1300 LT at Ajaccio and at 1300, 1400 or 1500 LT at Corte as well as from profiles measured by microwave radiometer (MWR) at Corte at 1300, 1400 and 1500 LT. The dashed lines indicate the thresholds for the TT index and RH and the respective categories I to IV (Sect. 1) are indicated in the corners.

about 57 and 58 °C, days with and without deep convection occurred both. When using profiles measured at Corte and applying the same classification as for Ajaccio, deep convection mostly occurred on days of category I, i.e. under conditions favourable for the evolution of deep convection.

On 25 August, the conditions in the pre-convective environment were assigned to category IV at Ajaccio and category I at Corte. To explain the differences of convection-related parameters between the two stations, the temporal evolution of the TT index and RH at Corte was investigated. The TT index at Corte at 1000 LT was similar to that at Ajaccio at 1300 LT, i.e. they both belonged to category IV (Fig. 6.21). After 1000 LT, the TT index and RH increased significantly at Corte, which was attributed to the evolution of a deep mountain ABL, and reached category I after 1300 LT. In mid-afternoon, the TT index was as high as

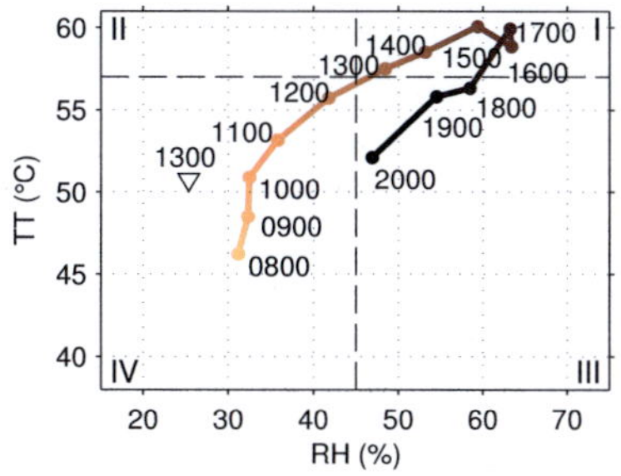

Figure 6.21.: TT index and RH at Corte (colour-coded) and at Ajaccio (black triangle) at 25 August. The parameters are derived from profiles measured by a radiosonde at Ajaccio and by a microwave radiometer at Corte. The numbers in the diagram denote time in LT. The dashed lines indicate the thresholds for the TT index and RH and the respective categories I to IV (Sect. 1) are indicated in the corners.

60 °C and RH varied around 60 %. In this environment deep convection evolved over the mountains in the north-eastern part of the island after around 1600 LT. This example well illustrates why deep convection was able to develop above the island despite unfavourable conditions on the coast.

6.3. Discussion

The combination of comprehensive measurements on the Corsican Island allowed to capture the complete transportation paths of water vapour, in contrast to most previous studies, where only parts of the transport processes or the final impact on the humidity distribution were observed (e.g. Ohtani, 2001; Henne et al., 2004). Based on the analysis of data from the whole field campaign and from individual case studies schematic diagrams were developed to summarize the characteristic features of the transportation paths of water vapour on days with undisturbed mountain ABL evolutions (Figs. 6.22a, b and c). Before the evolution of thermally driven circulations in the early morning, the water-vapour distribution above the mountain ridges was horizontally homogeneous in the idealized case without any large-scale humidity gradients. The IWV mainly depended on station height.

During the morning, thermally driven circulations developed (Fig. 6.22a). A sea breeze, which was detected by near-surface measurements, sodar and radiosondes on the east coast, inland transported maritime air. On the west coast, the regular evolution of a sea breeze was captured by near-surface measurements and radiosoundings. The humidity in this layer was often lower than on the east coast and it is speculated that this may have been related to different coast characteristics. Because the sea-breeze layer depth typically did not exceed a couple of hundred metres, it affected the IWV evolution on the coasts only little, causing weak IWV increases, if at all. Upslope winds, which were measured by a scanning wind lidar and at a surface station at a slope, could carry moist air from the valleys up the slopes, i.e. topographic venting could occur. This frequently caused a decrease of IWV in the valleys in the morning. Horizontal humidity differences formed with higher humidity values above the slopes and ridges, which were detected in microwave-radiometer scans and aircraft measurements and confirmed the existence of topographic venting of moist air. The transport of water vapour was associated with the evolution of a deep mountain ABL above the mountain ridges. Its depth varied significantly, depending on the near-surface temperature, i.e. the season of the year, and on the atmospheric stratification above the ridges. Higher near-surface temperatures and weaker stable stratification favoured the evolution of a deep mountain ABL.

Water vapour was also inland transported from the coast along the Tavignano Valley, which was attributed to the superposition of upvalley wind and sea breeze, resulting in a combined upvalley-wind and sea-breeze layer. This inland transport was detected by surface stations, microwave radiometers and radiosondes. The moist air in the upvalley-wind layer arrived at Corte in the centre of the island in late morning or early afternoon, contributing to an increase of IWV. Once the water vapour from the valley was transported with the upslope winds, the horizontal humidity differences above the ridges enhanced. The evolution of humidity differences allowed the advection of water vapour downstream of the mountain ridges, i.e. advective venting was present. Radiosondes measured elevated humidity layers and GPS and microwave radiometers detected an IWV increase over and downstream of the mountain ridges in the

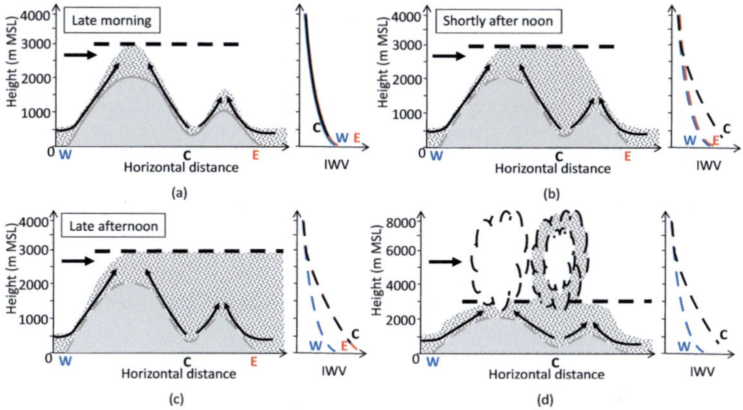

Figure 6.22.: Schematic diagrams illustrating the transportation paths of water vapour on days with undisturbed mountain ABL evolutions (a), (b) and (c) and on days with isolated deep convection (d). The water-vapour distribution under westerly upstream flow is indicated by the dotted areas along a west-east cross section through the island orography in the panels on the left. The conditions in late morning (a), shortly after noon (b) and late afternoon (c) are displayed. The thick dashed line indicates the mountain ABL height. The panels on the right show the schematic IWV profiles on the west coast (W), in the centre (C) and at and east coast (E) of the island. In (d), a dissolving deep convective cell that developed above the mountain gets downstream advected and affects the water-vapour distribution in and above the mountain ABL.

interior of the island shortly after noon (Fig. 6.22b). Vertical coupling of surface-based convective cells to elevated updraughts, observed in the Tavignano Valley in the interior of the island (Sect. 5.1), could also provide an effective vertical upward transport of water vapour, possibly enhancing the moistening of a deep mountain ABL.

Although advective venting of moist air from the lower mountain ridge in the north-eastern part of the island towards the east coast could have occurred, this process was not visible in the measurements on the east coast. The advective venting on the east coast rather occurred in the same layers as at Corte in the centre of the island. The impact of advective venting on the downstream coast at San Giuliano was typically delayed by a few hours compared to Corte. This was attributed to the spatial distance between the main mountain ridge and the east coast. Once the elevated moist layers that formed above the main mountain ridge arrived on the downstream coast, the IWV increased and the water-vapour distribution resembled the one at Corte in late afternoon (Fig. 6.22c).

Because mainly westerly large-scale upstream conditions prevailed during the measurement period, the east coast was frequently affected by advective venting with the mean large-scale wind and the evolution of a deep mountain ABL. It was found that the local flow on the west coast regularly exhibited an easterly component, even with a large-scale westerly wind component. This might have been related to the formation of a return flow. Similar flow features were found on the east coast on the few days with large-scale conditions with an easterly wind component. Although the easterly return flow component on the west coast or the westerly return flow component on the east coast would allow advective venting of moist air from the mountain ridges, the IWV only underwent a diurnal cycle when the large-scale upstream conditions exhibited an easterly or westerly component, respectively. This indicates that the duration and vertical extent of advective venting with the return flows on the coasts was not sufficient to show in the IWV evolution. Overall, the IWV evolution on a downstream coast was mainly caused by the transport of water vapour in elevated layers with the large-scale wind resulting in the evolution of a deep mountain ABL. The impact of advective venting on the downstream coast could explain the observed mean assymetric IWV distribution between the west and east coast.

While the maxima of IWV over the mountains (e.g. Wu et al., 2003) and on the downstream side (e.g. Iwasaki and Miki, 2001) in late afternoon were reported before, the diurnal cycle of IWV over the Tavignano Valley differed from previous studies, in which a decrease of IWV during the day in a valley was described (e.g. Kuwagata et al., 2001). Instead, Corte in the Tavignano Valley experienced the highest mean IWV value and the largest diurnal IWV variation from all stations located in the interior of the island. This was mainly caused by the superposition of three processes: in the morning before the sea-breeze front arrived at Corte, topographic venting might transport moist air from the valley centre up the slopes, frequently causing a decrease of IWV. Typically in the early afternoon, the humidity in the Tavignano Valley increased in the combined upvalley-wind and sea-breeze layer and in elevated layers above the valley, affected by advective venting, leading to a strong rise of IWV. The impact of advective venting on the IWV evolution at Corte was almost independent of the upstream wind direction, probably because Corte was surrounded by mountain ridges.

When isolated deep convection developed over the mountains and was downstream advected over Corte, the humidity increased in layers far above the mountain ABL and the IWV rose by several kg m^{-2} (Fig. 6.22d). The humidity increase was presumably caused by vertical motions within the convective cell, which transported water vapour upwards causing an exchange between the mountain ABL and the free atmosphere. Once the cloud dissolved, evaporation additionally contributed to an increase of humidity in the free atmosphere. In addition to convection and thermally driven circulations in the mountain ABL, vertical transport in a deep convective cell added to the transportation paths of water vapour.

When the background flow impacted the mountain ABL, the typical diurnal evolution of humidity and its spatial distribution were modified (Sec. 6.1.3). During one case study, the IWV significantly decreased at Corte because dry air from the free atmosphere intruded the mountain ABL. Observations at Corte indicated the generation of water-vapour variability on a small scale around Corte, i.e. on a scale of a few tenth of kilometres, which lasted for about one hour. At the other GPS stations in the interior of the island, the IWV still revealed the impact of a mountain ABL, which confirmed the confined spatial scale of the event. In another case study, the downward transport of warm and dry air from the free atmosphere downstream of the mountain ridges generated horizontal humidity variability on the scale of the island, which lasted for several hours. During this event no impact of a mountain ABL at any of the GPS stations in the interior of the island was observed. The data indicate that the intrusion of warm and dry air from the free atmosphere into the mountain ABL downstream of a mountain ridge was an important process producing water-vapour variability on different time and length scales, which were primarily controlled from above and not from the surface. The IWV data from spatially distributed GPS stations proved to be extremely useful for analysing spatial variabilities of water vapour and the associated mountain ABL over complex terrain.

The observations during the HyMeX field campaign confirmed the assumption from the statistical analysis (Sect. 1) that the conditions at Ajaccio were not representative for the environment over the mountains in which deep convection developed, i.e. deep convection occurred rather independent of mid- and low-level humidity and static stability at Ajaccio (Sect. 6.2). In fact, the conditions at Corte were much more meaningful for describing the atmospheric conditions over the mountains. Even when the conditions for the evolution of deep convection were unfavourable on the coast, the evolution of a deep mountain ABL at Corte strongly favoured the conditions in the pre-convective environment in which deep convection finally evolved. It is concluded that radiosoundings at Corte allowed to distinguish and predict days with deep convection and the soundings therefore are representative for the pre-convective environment. This means that observations with a spatial resolution on the mesoscale are necessary to properly relate deep convection over complex terrain to the pre-convective environment. The threshold used for the TT index agreed with values applied in previous studies (Ducrocq et al., 1998). As far as is known, RH, i.e. the humidity content between 850 and 700 hPa, has not been used to characterize the conditions in the pre-convective environment. Thus, its threshold cannot be compared with values for other areas. The observations suggest that RH is a meaningful parameter when characterizing the pre-convective environment over the Corsican Island by means of radiosonde profiles obtained in the island centre. However, the limited number of cases with deep convection during the HyMeX field campaign did not allow to statistically assess the skill of the TT index and RH at Corte for predicting isolated deep convection.

For a small range of TT index and RH values at Corte, days with as well as without deep convection occurred. Possible reasons for this could be the uncertainty range of the TT index and RH values, based on the measurement accuracy of the temperature and humidity data, but also that the atmospheric conditions at

Corte still did not represent the pre-convective environment over the interior of the island where deep convection evolved. For example, the initiation of deep convection more likely takes place above the mountain ridges, where thermally driven circulations provide a trigger mechanism, i.e. several kilometres away from Corte. The latter possible reason implies that the evolution of convection sometimes depends on spatial atmospheric variabilities on an even smaller scale than resolved by the installed network on the Corsican Island.

7. Summary and Conclusions

7.1. Summary

During late summer and autumn, isolated deep convection frequently develops over the mountainous island of Corsica in the western Mediterranean Sea. A statistical analysis of the pre-convective environment on the west coast for a 10-year time period indicated that deep convection over the island in many times evolved despite strong stable stratification and low humidity in the middle atmosphere, i.e. conditions not at all favourable for the occurrence of deep convection. Possible reasons for that are mesoscale spatial variabilities of water vapour and atmospheric stratification, which are considered as decisive for the evolution of deep convection. Over complex terrain, convection and mesoscale thermally driven circulations develop under fair weather conditions during daytime. The co-existence and superposition of the associated various transport processes produce a diurnal variation of atmospheric variables in a deep atmospheric layer, named mountain ABL. Nevertheless, the superposition of relevant transport processes and the evolution of the mountain ABL and its impact on the pre-convective atmospheric environment are still not completely understood, because observations over complex terrain especially on the transition between the turbulent scale and the mesocale are rare. Thus, this thesis focused on identifying multi-scale transport processes in the mountain ABL over the Corsican Island by means of comprehensive measurements. Furthermore, the influence of these processes on the spatial variabilities of water vapour, convection-related parameters and the evolution of deep convection was assessed and the benefit of integrated measurement systems was investigated.

The used data were mainly gathered with the mobile observation platform KITcube on the Corsican Island during the HyMeX field campaign in late summer and autumn 2012. The island's dimension amounts to about 180 km in north-south and 80 km in west-east direction. It features a main north-northwest to south-southeast oriented mountain ridge with more than 20 peaks higher than 2000 m MSL, and another somewhat lower mountainous area in the north-western part of the island. In between these areas, the Tavignano Valley runs from the east coast to Corte in the centre. The KITcube measurement systems were distributed over two main deployment sites: one station was located at Corte in the centre of the island in the Tavignano Valley and another one on the east coast at San Giuliano. Corte was equipped with e.g. various in-situ devices to measure mean and turbulent variables near the surface; a radiosonde system to obtain vertical profiles of temperature, humidity and wind; a scanning microwave radiometer, which e.g. provides spatial information about the water-vapour distribution; and two wind lidars and a cloud radar for radial velocity measurements under cloud-free and cloudy conditions. Its location in the Tavignano Valley east of the main mountain ridge enabled to observe the mountain ABL and convection-layer evolution in the valley. Three mobile towers measuring near-surface mean and turbulent variables collected data along the axis of the Tavignano Valley. Surface stations, a radiosonde system and microwave radiomter at San Giuliano covered the coastal to maritime atmospheric conditions on the east coast, while operationally launched radiosondes and an automatic weather station at Ajaccio provided some information about the atmospheric characteristics on the west coast. A permanently operated GPS network, consisting of 16 stations distributed over the island, delivered information about the spatial IWV distribution. Aircraft data from west-east flight tracks across the island at different levels yielded spatial variability of mean and turbulent variables above the mountain ridges on some days.

In order to detect the top of the mountain ABL and the part of the mountain ABL which was affected by surface-based, buoyancy-driven turbulent mixing objective methods were developed and implemented. Standard methods to detect the CBL height were applied to data measured at Corte over complex terrain as well as over homogeneous terrain. While the CBL height over homogeneous terrain determined with different algorithms gave similar results and also agreed with the layer in which atmospheric variables showed a diurnal cycle, large variations in the differently detected CBL heights occurred over complex terrain. Because the variance of the vertical wind speed measured by wind lidars was the only available variable that directly represented turbulence, it was applied to detect the top of the layer dominated by buoyancy-driven turbulent mixing in the Tavignano Valley. To assure that the turbulent mixing was also surface-based, the skewness of the vertical wind speed was used as an additional criterion. When turbulent mixing is mainly surface-based and buoyancy-driven in an atmospheric layer, the skewness is generally positive throughout that layer. The height of the layer dominated by convection was detected as the level above which the variance fell below a certain threshold and the skewness was mainly positive below this level. As the vertical structure of temperature and humidity in this layer over complex terrain did not resemble the structure of a classical CBL over homogeneous terrain, this layer was termed convection layer instead of CBL to avoid confusion. The mountain ABL detection method was based on humidity profiles measured by radiosondes. The level up to which a significant diurnal increase of specific humidity occurred between subsequent profiles was identified as mountain ABL top. Due to the additional impact of advection related to thermally driven circulations, the height of the mountain ABL and convection layer in the Tavignano Valley typically differed in the course of the day.

Motivated by the results of the statistical analysis several scientific objectives were formulated, which were addressed in detail by means of case studies as well as long-term observations during the field campaign.

1. The first main objective was to identify the typical undisturbed evolution of the mountain ABL under fair weather conditions and to determine the processes which were responsible for its evolution. Based on data from 19 August 2012, the characteristics of a mountain ABL with an undisturbed evolution and the processes governing this evolution were identified and illustrated in a conceptual model focusing on Corte.

   A combined upvalley-wind and sea-breeze layer established in the Tavignano Valley. The arrival of the sea-breeze front was derived from a strong humidity increase in the upvalley-wind layer at Corte at around noon. Strong mean subsidence in the order of -1 m s^{-1} lasting for several hours occurred in an atmospheric layer of several hundred metres depth within the mountain ABL above the valley centre. The estimation of the mass balance for an air volume centred at Corte revealed that the subsidence was likely induced by low-level divergence mostly due to upslope flows. There was evidence that the compensating subsidence reduced the growth of the convection layer, but at the same time contributed to the breakup of the nocturnal surface inversion in the valley by warming from above. In the afternoon, warming due to mean subsidence and horizontal advection within the stably stratified combined upvalley-wind and sea-breeze layer probably contributed to the stabilization in parts of the convection layer despite vertical mixing due to convection.

Above the long-lasting subsidence layer in the valley elevated vertical motions were found, consisting of alternating up- and downdraughts. When the convection layer in the valley centre reached the bottom of the layer with the elevated vertical motions, updraughts in the two layers vertically coupled and cells reached from the surface up to top of the layer with elevated vertical motions, resulting in a strong growth of the convection-layer depth. Simultaneously, the humidity and aerosol content rapidly increased in the layer with the elevated vertical motions. This suggests that cell coupling caused an effective vertical transport, which resulted from the superposition of turbulent and mesoscale processes. Thus, a dependency of the convection-layer growth in the valley on various processes was evident: subsidence reduced its growth, while a previous labilisation due to advection or vertical cell coupling enhanced it.

Horizontal humidity inhomogeneities formed in elevated layers with higher humidity values above the mountain ridges. These were likely related to topographic venting, i.e. an upward transport of moist air with upslope winds, and by convection above the ridges. Simultaneously, a complex flow structure evolved at Corte in the Tavignano Valley during the day and three different wind layers, i.e. the upvalley wind and two elevated wind layers, were identified, all showing a diurnal variation and modifying the large-scale north-easterly flow. Likewise, an elevated layer with a westerly wind component evolved on the east coast in the afternoon within the large-scale flow. The elevated wind layers might have been the return flows of thermally driven circulations. Depending on the wind direction, moister air from above the ridges was transported downstream with the mean wind, i.e. advective venting occurred, and explained the formation of elevated moist layers and a deep mountain ABL at Corte and on the coasts.

The advective venting of moist air resulted in the evolution of a deep mountain ABL which extended up to about 3200 m MSL, i.e. about 2800 m above the floor of the Tavignano Valley at Corte. Because the mountain ABL resulted from convection as well as thermally driven processes, it was considerably deeper than the convection layer in the valley and also deeper than a CBL that would evolve over homogeneous terrain under equal surface forcing. The estimated level of neutral buoyancy of an air parcel rising from the mountain ridges well agreed with the mountain ABL height and it was concluded that the mountain ABL height depends on mountain-ridge height, near-surface temperature and atmospheric stratification.

2. The second main objective focused on how the mountain ABL evolution was influenced by the background flow. To address this objective, two case studies were carried out, when the evolution of the mountain ABL was interrupted (02 October 2012) or inhibited (09 October 2012). As clear skies and high surface sensible heat fluxes would allow for convection and thermally driven circulations on both days, it was expected that the large-scale background flow caused the different mountain ABL evolutions. Both analysed days were characterized by a large-scale westerly wind component so that Corte was downstream of the main mountain ridge. However, the upstream wind was more than twice as strong on 09 October 2012 than on 02 October 2012.

   On 02 October 2012, a convection layer, a south-easterly upvalley wind layer and a mountain ABL established at Corte during the day. In the afternoon, a layer developed, in which the potential temperature and specific humidity were about constant with height and which was deeper than the convection layer. Within this layer, warm and dry air with a westerly wind component, accompanied by strong upward motions, gradually replaced the air of the mountain ABL. This air mass approached Corte from the west and originated from a layer in the free atmosphere around ridge height. Between the intruding air mass with a westerly wind component and the south-easterly upvalley-wind layer a convergence zone developed, which was located right at Corte and resulted in strong upward motions. The typical mountain ABL evolution with thermally driven circulations and convection was interrupted at Corte for about one hour.

Already in the morning of 09 October 2012 a warm and dry air mass with a westerly wind component, which orignated from the free atmosphere, intruded the atmosphere in the Tavignano Valley at Corte. This was associated with downward motions. Once the nocturnal surface inversion was eroded, this air mass reached the ground, causing a sudden change of near-surface conditions: the temperature increased, humidity and particle density decreased, the variance of the vertical wind speed increased and the wind gained a westerly large-scale component and accelerated. As similar changes in near-surface conditions occurred at the same time at Pont génois, which is located south-east of Corte about 15 kilometre further down the valley, it was concluded that the intruding air mass extended far along the valley. Because a typical upvalley wind evolved closer to the mouth of the Tavignano Valley during the day, horizontal convergence existed between Pont génois and the valley mouth. The intrusion of free-atmosphere air at Corte and Pont génois lasted for several hours and inhibited the evolution of thermally driven circulations and a mountain ABL in this section of the valley. In the afternoon, the upvalley wind, which prevailed further down the valley all day long, pushed up the valley and gradually replaced the intruding air mass.

The conditions at Corte presumably depended on the location of the convergence zone which was generated between the intruding warm and dry air mass with a westerly wind component and the cool and moist upvalley-wind layer. It occurred around Corte producing upward motions at the station on 02 October 2012, while it existed much further downstream of Corte on 09 October 2012 resulting in strong downward motions at Corte. To explain the intrusion of the warm and dry air mass, several possible mechanisms were investigated, i.e. turbulent transport, large-amplitude lee waves or downslope windstorms, and the likelihood of their occurrence was discussed for these two days.

3. The third scientific objective was to investigate how the evolution of the mountain ABL affected the temporal and spatial variabilities of water vapour and atmospheric stratification over the Corsican Island and to study whether information on a higher spatial resolution allowed a better understanding of the evolution of deep convection. This objective was addressed based on data from the three-month measurement period and from individual case studies.

   The atmospheric water-vapour distribution was closely related to the evolution of the mountain ABL. Large differences in the IWV evolution occurred on days with undisturbed, interrupted and inhibited mountain ABL evolutions. An undisturbed evolution of a deep mountain ABL was typically accompanied by an increase of IWV shortly after noon, maximum IWV values in the late afternoon or early evening and a decrease afterwards. Besides the station height, the IWV depends on whether the atmosphere above the station was moister in regard to the reference atmosphere, i.e. whether it was affected by a deep mountain ABL. From the whole HyMeX field campaign period, days where selected when the temporal IWV evolution was presumably primarily controlled by processes in the mountain ABL. For these days, the mean spatial distribution of IWV over the Corsican Island was analysed by means of GPS measurements. The most interesting findings are that the mean IWV in the afternoon was generally lower on the west coast than on the east coast at the same height; the IWV at all stations in the interior of the island was significantly higher than in the reference atmosphere at the corresponding heights, i.e. they were all influenced by the evolution of a deep mountain ABL; and the strongest diurnal cycle of IWV occurred at Corte.

The layers contributing to the IWV evolution on the coasts and in the interior of the island at Corte were analysed exemplarily for two days with undisturbed mountain ABL evolutions, one in August and one in October. The most significant transport processes were found to be advection of water vapour in the combined upvalley-wind and sea-breeze layer in the Tavignano Valley and topographic and advective venting. While the elevated horizontal humidity gradients, necessary for advective venting, were related to the orography, the transport of water vapour mainly occurred with the large-scale wind. These processes resulted in a large spatial variability of water vapour between the interior of the island and the upstream and downstream coasts. With mainly westerly large-scale upstream conditions, the processes explained the higher mean IWV on the east coast compared to the west coast. As Corte in the Tavignano Valley was affected by the water-vapour transport in the combined upvalley-wind and sea-breeze layer, as well as by a transport of moist air from the surrounding mountain ridges due to advective venting, the diurnal cycle of IWV was most pronounced at this station. The diurnal cycle of IWV revealed a seasonal dependency with higher IWV amplitudes in August than in September or October. This was attributed to deeper mountain ABL depths in August, which were mainly related to higher near-surface temperatures. When intrusions of warm and dry air from the free atmosphere interrupted or inhibited the mountain ABL evolution, the diurnal cycle of IWV was also interrupted or not evident in the respective areas.

To characterize the pre-convective environment over the Corsican Island, the TT index accounting for static stability and low-level humidity and the relative humidity (RH) averaged between 850 and 700 hPa characterizing the mid-level humidity turned out to be suitable. The great advantage of these convection-related parameters was, compared to CIN and CAPE, that they were defined for each sounding and could also be calculated from microwave radiometer measurements. The evolution of a mountain ABL at Corte typically led to an increasing TT index mainly due to a warming and moistening of the lower troposphere and an increasing RH mostly because of an increasing humidity in mid-levels, i.e. the conditions became more favourable for the subsequent evolution of deep convection. An interruption or inhibition of the mountain ABL evolution at Corte due to an intrusion of dry and warm air from the free atmosphere led to low RH and TT values. When a deep mountain ABL evolved over the Corsican Island, the atmospheric conditions in the interior of the island were much more favourable for the development of deep convection than on the west coast. A comparison of days with and without isolated deep convection showed that the occurrence of deep convection over the island was rather independent of the atmospheric conditions on the west coast, as was expected from the statistical analysis. Deep convection occurred over the island for values of the TT index and RH on the west coast indicating favourable as well as unfavourable conditions for deep convection. In contrast, the TT index and RH at Corte were mainly higher than the thresholds on days with deep convection, which means that Corte is a representative site to predict deep convection over the island. Thus, spatial variabilities of convection-related parameters occurred on the mesoscale due to the evolution of the mountain ABL, which explains why deep convection evolves over the Corsican Island under unfavourable conditions on the west coast.

7.2. Overall Conclusions

In this thesis, the superposition of multi-scale boundary-layer processes and their importance in producing water-vapour variability was proved and the transporation paths of water vapour were identified by means of comprehensive measurements. Due to their high complexity, simultaneously observing all processes from the turbulent scale to the mesoscale is a great challenge. However, measuring relevant processes involved is necessary to improve the understanding of the whole process chain and transportation paths. From the KITcube measurements on the Corsican Island during HyMeX new aspects were gained, which are highlighted in the following. The combination of such a large number of in-situ and remote sensing instruments available on the island, particularly at

Corte, was quite unique. The integration of different instruments, the implementation of the designed measurement configuration and coordinated scan strategies successfully allowed to simultaneously capture relevant processes and enhanced the process understanding. For example, while wind-lidar scans revealed the existence of upslope winds, parallel microwave-radiometer scans indicated the accumulation of humidity above the higher terrain. This confirms water-vapour transport with the upslope winds. Furthermore, the cloud radar provided additional information about vertical motions in layers where the aerosol content was obviously not sufficient for the wind lidar to return valid data. This contributes to a better understanding of the prevailing vertical transport processes.

Several previously unreported boundary-layer processes were observed at Corte: the combination of radiosondes, a wind lidar, a cloud radar and a microwave radiometer provided experimental evidence of a vertical transport process in a valley when surface-based convective cells coupled to elevated updraughts. The intrusion of dry and warm air with an upstream wind component from the free atmosphere down to the surface in a downstream valley generated mesoscale water-vapour variability, which was controlled from above and not from the surface. Three independent instruments provided evidence for strong, compensating subsidence in a valley. As such strong subsidence has not been observed by continuous remote sensing at a valley site before, this suggests that its impact on the evolution of the mountain ABL and especially on the depth of the convection layer has possibly been underestimated so far.

The complex terrain of the Corsican Island and the humid environment of the western Mediterranean provided ideal conditions to study the influence of a mountainous island on the evolution of deep convection. The impact of the terrain on deep convection could clearly be demonstrated: transport processes ranging from the turbulent to the mesoscale lead to the evolution of a mountain ABL and cause variabilities of humidity and convection-related parameters over the Corsican Island, which are decisive for the development of isolated deep convection. This means in order to properly relate deep convection to the pre-convective environment, the conditions, especially in low- and mid-levels, have to be known with high spatio-temporal resolution. As great parts of the Earth's surface consist of complex or mountainous rather than homogeneous terrain, comparable dependencies can be expected for most of those regions, too. Thus, the derived findings may give valuable hints for the interpretation of precipitation events assigned to mesoscale inhomogeneities.

Acknowledgement

I would like to thank my supervisor Prof. Christoph Kottmeier for his constant support and mentoring. I am also very grateful to PD Dr. Michael Kunz for co-supervising this thesis.

I am particularly thankful to Dr. Norbert Kalthoff for the excellent support and discussions, for always taking time to help me and for the opportunity to participate on various field campaigns during my time at IMK.

I am also grateful to Jan Handwerker, Martin Kohler, Andreas Wieser, Stephan Kraut, the technical staff and the whole IMK team for their efforts to deploy the KITcube on Corsica. Further, I thank Markus Ramatschi from GFZ Potsdam for installing the GPS stations on Corsica, Olivier Bock from IGN LAREG for providing the IWV data, Samiro Khodayar for her help in making the IWV data available, Michael Kunz for his help in accessing the lightning data, Katja Träumner for her help with the wind lidar data and Ulrich Löhnert from University of Cologne for providing retrievals and the postprocessing software for the microwave radiometers. I am also thankful to Rolf Hankers and the aircraft team from University of Braunschweig for performing the aircraft measurements during HyMeX and Stefan Heise and Jens Wickert from GFZ Potsdam and Bernhard Pospichal from University of Leipzig for loaning microwave radiometers.

Many thanks to my working group colleagues for their support and help. Especially, I would like to thank Vera Maurer for sharing the room with me, creating a pleasant atmosphere, letting fresh air in and taking care of my tea needs.

Also, I wish to thank Gabi Klinck for her support with the technical equipment, Rosalba Gräbner and Doris Stenschke for their help with administrative problems and Martina Zieger for proofreading my thesis. Many thanks to Tobias Maurer for his spontaneous help in fixing the cloud radar on Corsica.

Finally, I am deeply grateful to Sebastian, my family, particularly my mother, and friends for their patience and support.

A. Acronyms and Symbols

Acronyms

ABL	Atmospheric Boundary Layer
ACTIPLAN	company
AGL	Above Ground Level
AROME	Applications of Research to Operations at MEsoscale
AROME WMED	HyMeX-dedicated version of the French operational AROME system
BLIDS	Siemens lightning information service
CAPE	Convective Available Potential Energy
CBL	Convective Boundary Layer
CIN	Convective Inhibition
Δtrop	tropospheric delay
EL	Equilibrium Level
ENSG	École Nationale des Sciences Géographiques
EUCLID	European Cooperation for Lightning Detection
EUMETSAT	European Organisation for the Exploitation of Meteorological Satellites
GFZ	Helmholtz-Zentrum Potsdam, Deutsches GeoForschungsZentrum
GPS	Global Positioning System
HD(CP)^2	High Definition of Clouds and Precipitation for Climate Prediction
HOPE	HD(CP)^2 Observational Prototype Experiment
HyMeX	Hydrological cycle in the Mediterranean eXperiment
IGN	Institut Géographique National
IOP	Intense Observation Period
IWV	Integrated Water Vapour

LAREG	LAboratoire de Recherche en Géodésie
LCL	Lifting Condensation Level
LFC	Level of Free Convection
lidar	light detection and ranging
LT	Local Time (UTC plus 2 hours)
METEO FRANCE	french weather service
MSG	Meteosat Second Generation
MSL	above Mean Sea Level
MWR	microwave radiometer
NBL	Nocturnal Boundary Layer
radar	radio detection and ranging
RH	mean relative humidity averaged between 850 and 600 hPa
RS	radiosonde
SNR	Signal-to-Noise Ratio
SOP	Special Observation Period
TAF	Topographic Amplification Factor
TKE	Turbulent Kinetic Energy
TT	Total Totals index
UTC	coordinated universal time
ZHD	Zenith dry or Hydrostatic Delay
ZTD	Zenith Total Delay
ZWD	Zenith Wet Delay

Symbols

α	absorption coefficient
κ	conversion factor
λ	wavelength
$\lambda_{\min}$	smallest possible wavelength for steady waves
ν_q	kinematic molecular diffusivity for water vapour
ϕ	azimuth angle
φ	geographic latitude
ρ_d	density of dry air
$\overline{\rho}_d$	mean density of dry air
ρ_m	density of moist air
ρ_p	particle density
ρ_v	water-vapour density
σ_i	backscatter cross section

σ_u^2	variance of the wind component in x-direction
σ_v^2	variance of the wind component in y-direction
σ_w^2	variance of the wind component in z-direction
τ	optical depth of the atmosphere
Θ	potential temperature
$\overline{\Theta}$	mean potential temperature
$\overline{\Theta}_0$	near-surface mean potential temperature
Θ_{m}	potential temperature of a lower layer in hydraulic flow theory
Ω	intrinsic angular frequency
ω	angular frequency in a fixed reference frame
A	complex wave amplitude
a	scalar
ac	aerosol content
B	complex wave amplitude
b	scalar
B_f	Planck function
c	phase speed
c_0	speed of light in vacuum
c_p	specific heat at constant pressure for dry air
D	flow-layer depth
d	scalar
D_t	target diameter
D_{upstream}	far upstream flow-layer depth
E	latent heat flux
e	water vapour pressure
E_0	surface latent heat flux
$\vec{e}_\phi$	basis vector in cylinder coordinates
$\vec{e}_{r_c}$	basis vector in cylinder coordinates
e_s	saturation water vapour pressure
$\vec{e}_z$	basis vector in cylinder coordinates
f	frequency
F_c	flux through cylinder
Fr	Froude number for hydraulic flow theory
g	gravity acceleration
H	barrier height
h	Planck constant
H_0	surface sensible heat flux
h_0	height of the Earth's surface
H_c	cylinder height

h_c	height from cylinder bottom
h_t	height of the top of the atmosphere
I_f	radiant power per unit area and per unit solid angle within a specified frequency interval
k	wavenumber in x-direction
$\|K^2\|$	factor related to the composition-dependent index of refraction
k_B	Boltzmann constant
L	specific latent heat of vaporization
l	Scorer parameter
M	unit mass
m	wavenumber in z-direction
$\overline{M}$	mass of water vapour per unit volume and time, which is created by phase changes from liquid or solid
$\mathrm{m_{hyd}}$	mapping function
$\mathrm{m_{wet}}$	mapping function
N	Brunt-Väisäla frequency
n	number of azimuth angles during a scan
$\vec{n}$	unit normal vector in cylinder coordinates
p	air pressure
p_0	surface pressure
P_1	transmitted power
P_2	received power
q	specific humidity
$\overline{q}$	mean specific humidity
R	correlation coefficient
R_c	cylinder radius
r_c	distance from cylinder centre
R_d	gas constant for dry air
rh	relative humidity
Ri	gradient Richardson number
Ri_b	bulk Richardson number
Ri_c	critical Richardson number
r_v	water-vapour mixing ratio
rv	radial velocity
rv_{xy}	radial velocity projected on x-y plain
S	skewness of the wind component in z-direction
s	path of propagation
S_c	cylinder surface

S_f	source term in the radiative transfer equation
T	physical temperature
t	time
T_B	brightness temperature
T_d	dewpoint temperature
T_{m}	water-vapour weighted mean temperature in an atmospheric column
T_v	virtual temperature
U	horizontal upstream flow speed
u	flow speed in x-direction
$\overline{u}$	mean flow speed in x-direction
u'	flow speed fluctuation in x-direction
u_g	cross-shore wind component resulting from the large-scale pressure gradient
u_{ridge}	horizontal fluid speed at the mountain ridge
u_{sb}	horizontal wind speed in a sea breeze
u_{sbf}	propagation speed of a sea-breeze front
$\overline{v}$	mean flow speed in y-direction
v'	flow speed fluctuation in y-direction
$\vec{v}$	velocity vector
V_c	cylinder volume
w	flow speed in z-direction
$\overline{w}$	mean flow speed in z-direction
w'	flow speed fluctuation in z-direction
w_*	convective velocity
$\tilde{w}$	complex amplitude
w_0	amplitude
z	height
z_1	height of the well-mixed layer
z_2	height where the free atmosphere starts
z_i	CBL height

B. Tables and Figures

Table B.1.: Analysed days from the HyMeX field campaign 2012 with undisturbed, interrupted and inhibited mountain ABL evolutions as well as days with isolated deep convection.

Undisturbed mountain ABL evolution	Interrupted mountain ABL evolution	Inhibited mountain ABL evolution	Isolated deep convection
15 August 19 August 24 August 21 September 23 September 05 October	02 October	09 October	23 August 25 August 06 September 09 September

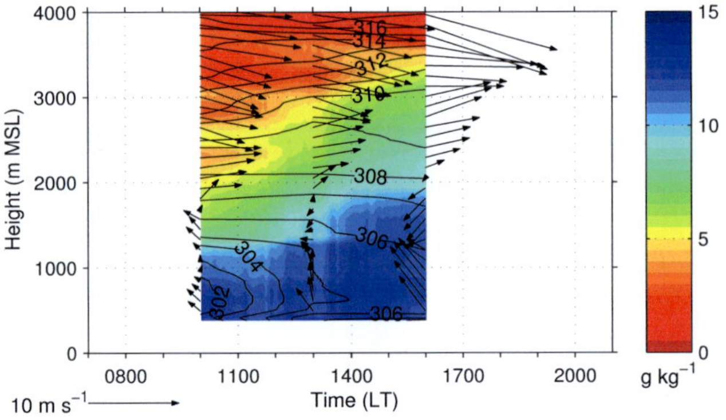

Figure B.1.: Specific humidity (colour-coded), potential temperature (black isolines) and horizontal wind vector (arrows) measured by radiosondes at Corte on 15 August.

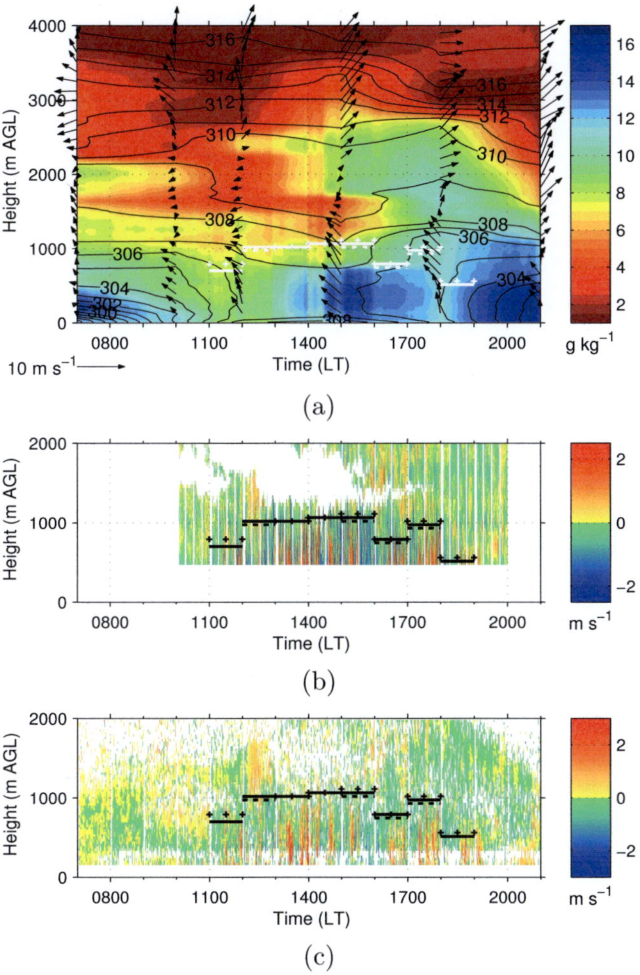

Figure B.2.: Specific humidity (colour-coded), potential temperature (black isolines) and horizontal wind vector (arrows) measured by radiosondes (a) and vertical wind speed measured by wind lidar (b) and cloud radar (c) at Corte on 24 August. Specific humidity is interpolated using microwave radiometer and radiosonde data and potential temperature is temporally linearly interpolated between the radiosoundings. The solid horizontal lines mark the variance- and skewness-based CBL depths and the thick dashed horizontal lines mark the lower and the crosses mark the upper boundaries of the uncertainty range if present.

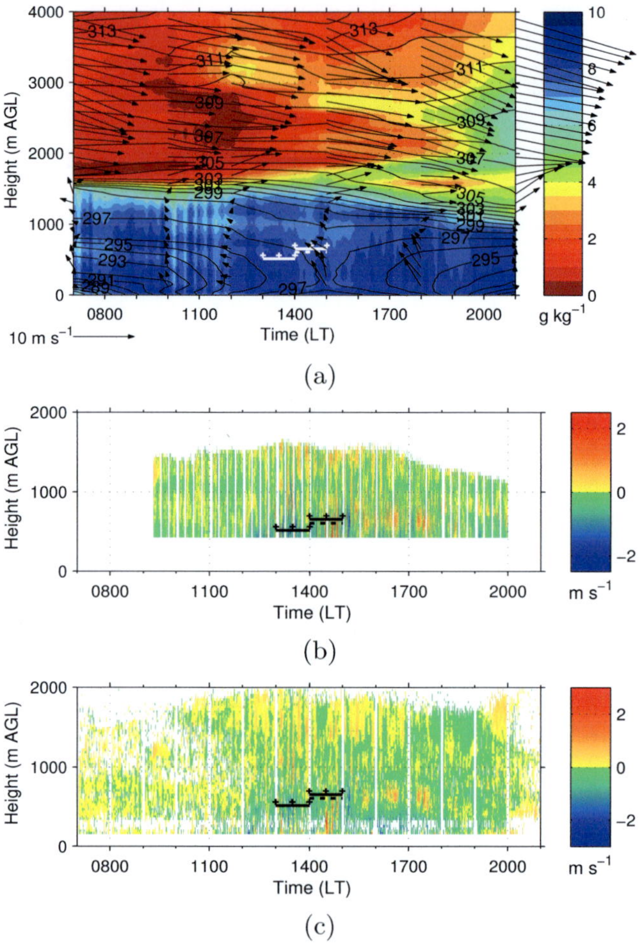

Figure B.3.: Specific humidity (colour-coded), potential temperature (black isolines) and horizontal wind vector (arrows) measured by radiosondes (a) and vertical wind speed measured by wind lidar (b) and cloud radar (c) at Corte on 21 September. Specific humidity is interpolated using microwave radiometer and radiosonde data and potential temperature is temporally linearly interpolated between the radiosoundings. The solid horizontal lines mark the variance- and skewness-based CBL depths and the thick dashed horizontal lines mark the lower and the crosses mark the upper boundaries of the uncertainty range if present.

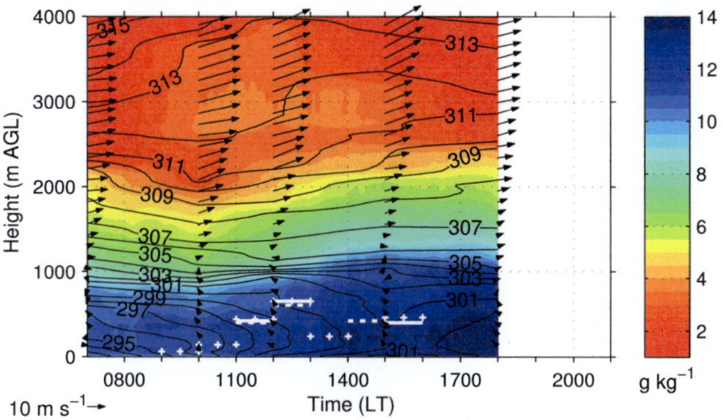

(a)

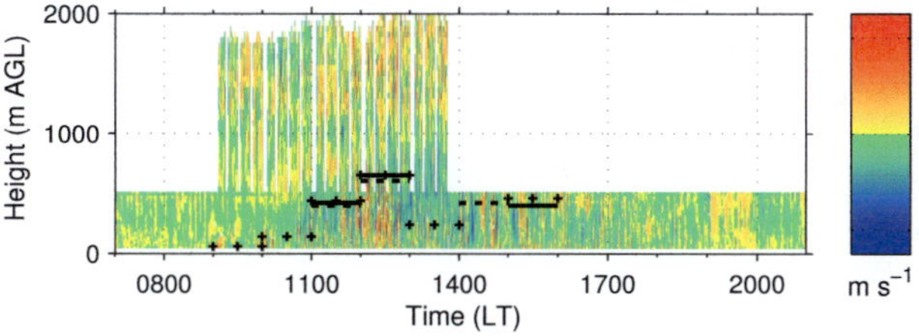

(b)

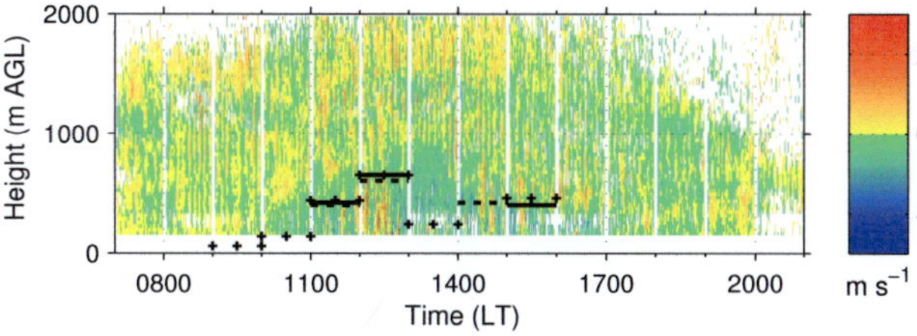

(c)

Figure B.4.: Specific humidity (colour-coded), potential temperature (black isolines) and horizontal wind vector (arrows) measured by radiosondes (a) and vertical wind speed measured by wind lidar (b) and cloud radar (c) at Corte on 23 September. Specific humidity is interpolated using microwave radiometer and radiosonde data and potential temperature is temporally linearly interpolated between the radiosoundings. The solid horizontal lines mark the variance- and skewness-based CBL depths and the thick dashed horizontal lines mark the lower and the crosses mark the upper boundaries of the uncertainty range if present.

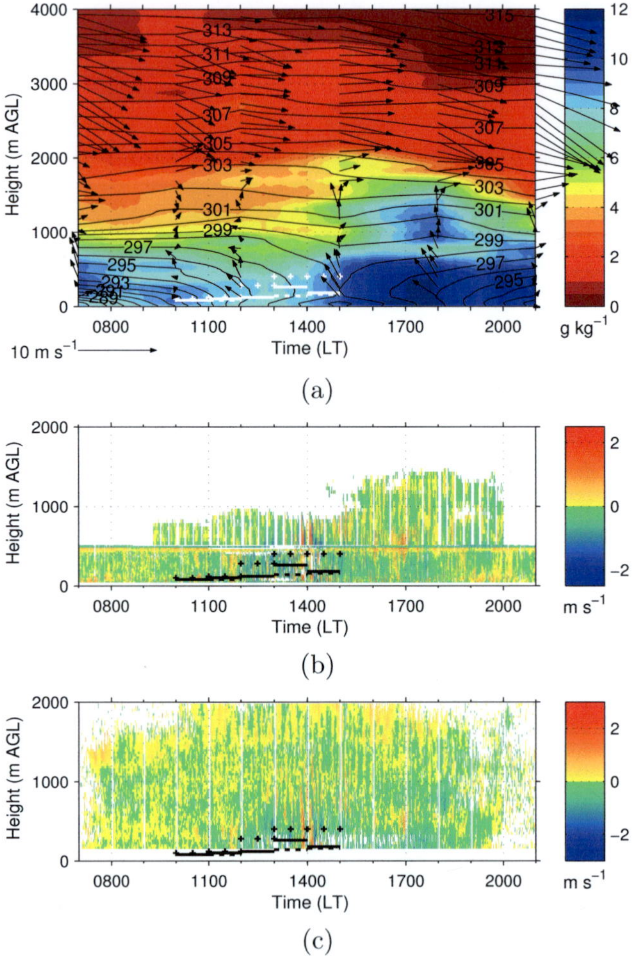

Figure B.5.: Specific humidity (colour-coded), potential temperature (black isolines) and horizontal wind vector (arrows) measured by radiosondes (a) and vertical wind speed measured by wind lidar (b) and cloud radar (c) at Corte on 05 October. Specific humidity is interpolated using microwave radiometer and radiosonde data and potential temperature is temporally linearly interpolated between the radiosoundings. The solid horizontal lines mark the variance- and skewness-based CBL depths and the thick dashed horizontal lines mark the lower and the crosses mark the upper boundaries of the uncertainty range if present.

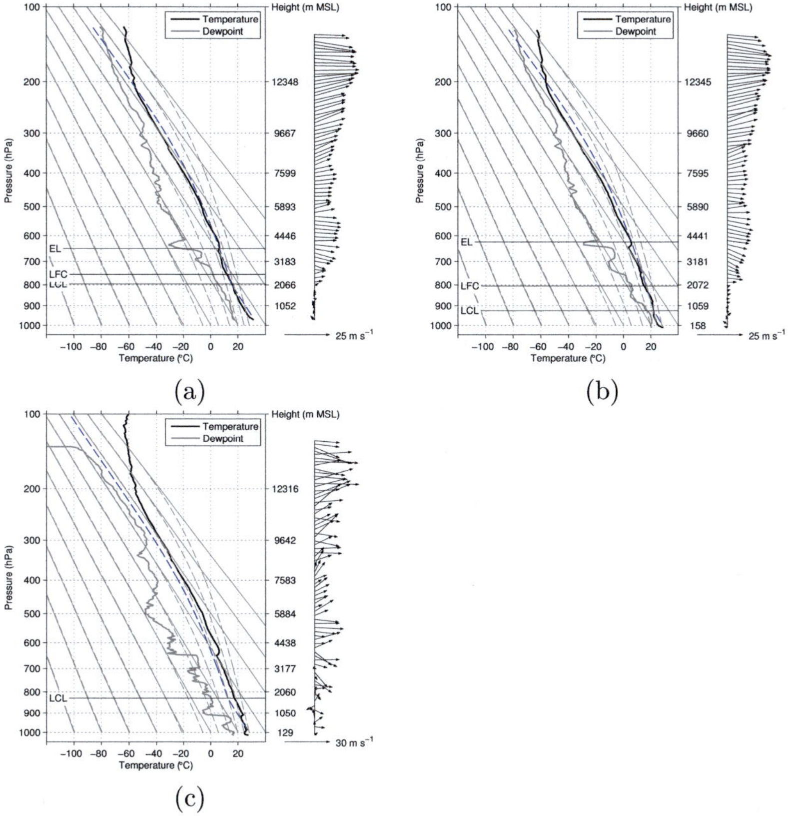

Figure B.6.: Radiosonde profiles of temperature, dewpoint and wind vector at 1300 LT at Corte (a), at 1300 LT at San Giuliano (b) and at 1300 LT at Ajaccio (c) on 15 August. The dashed blue lines indicate the virtual temperature of an air parcel with the mean layer temperature and humidity values of the lowest 50 hPa, which is lifted from the near the surface to its Lifting Condensation Level (LCL), Level of Free Convection (LFC) and Equilibrium Level (EL).

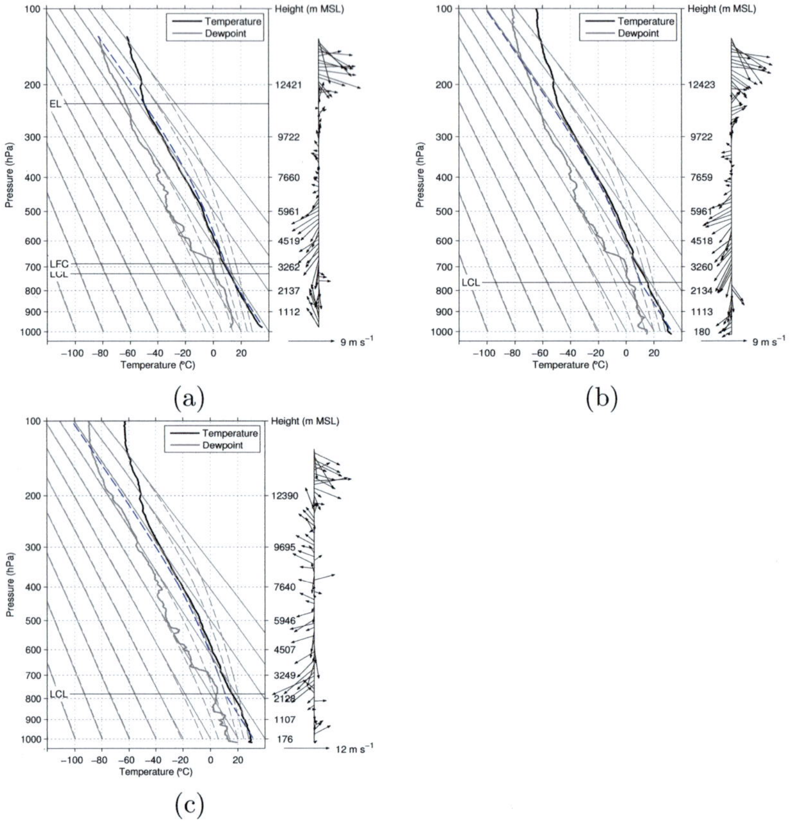

Figure B.7.: Radiosonde profiles of temperature, dewpoint and wind vector at 1400 LT at Corte (a), at 1400 LT at San Giuliano (b) and at 1300 LT at Ajaccio (c) on 19 August. The dashed blue lines indicate the virtual temperature of an air parcel with the mean layer temperature and humidity values of the lowest 50 hPa, which is lifted from the near the surface to its Lifting Condensation Level (LCL), Level of Free Convection (LFC) and Equilibrium Level (EL).

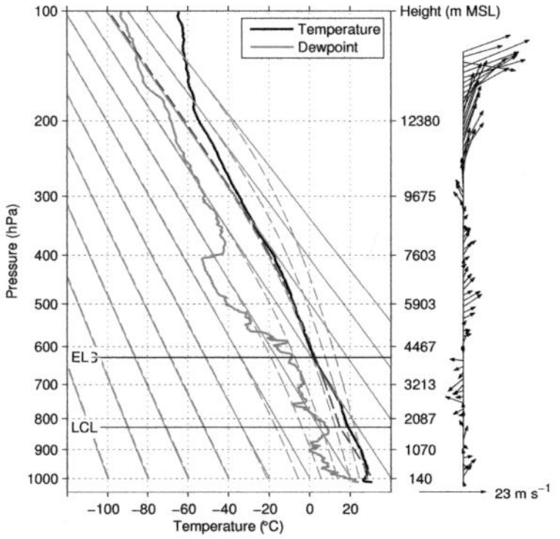

Figure B.8.: Radiosonde profiles of temperature, dewpoint and wind vector at 1300 LT at Ajaccio on 23 August. The dashed blue lines indicate the virtual temperature of an air parcel with the mean layer temperature and humidity values of the lowest 50 hPa, which is lifted from the near the surface to its Lifting Condensation Level (LCL), Level of Free Convection (LFC) and Equilibrium Level (EL).

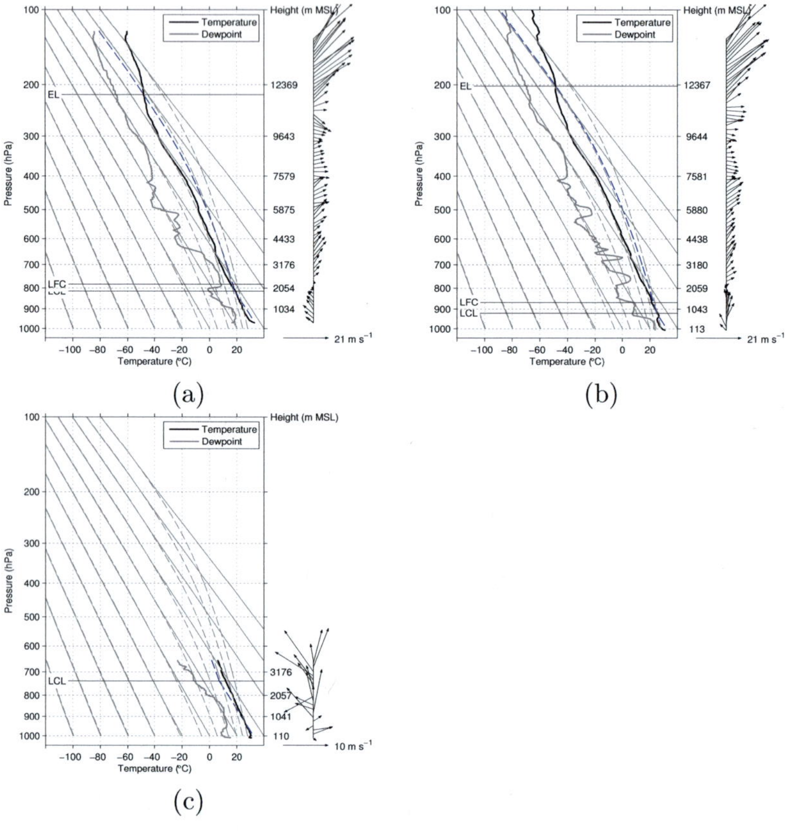

Figure B.9.: Radiosonde profiles of temperature, dewpoint and wind vector at 1500 LT at Corte (a), at 1500 LT at San Giuliano (b) and at 1300 LT at Ajaccio (c) on 24 August. The dashed blue lines indicate the virtual temperature of an air parcel with the mean layer temperature and humidity values of the lowest 50 hPa, which is lifted from the near the surface to its Lifting Condensation Level (LCL), Level of Free Convection (LFC) and Equilibrium Level (EL).

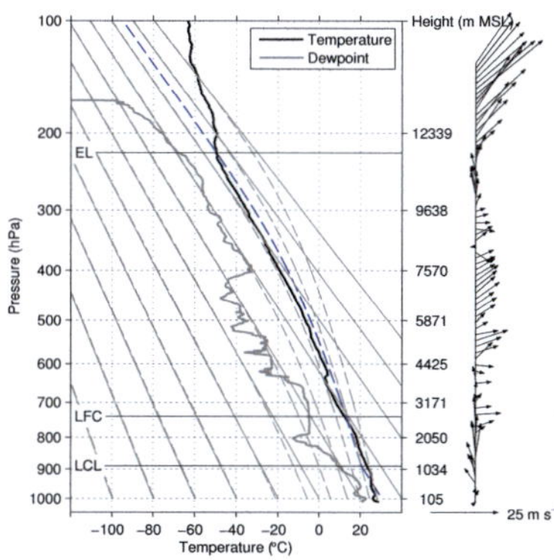

Figure B.10.: Radiosonde profiles of temperature, dewpoint and wind vector at 1300 LT at Ajaccio on 25 August. The dashed blue lines indicate the virtual temperature of an air parcel with the mean layer temperature and humidity values of the lowest 50 hPa, which is lifted from the near the surface to its Lifting Condensation Level (LCL), Level of Free Convection (LFC) and Equilibrium Level (EL).

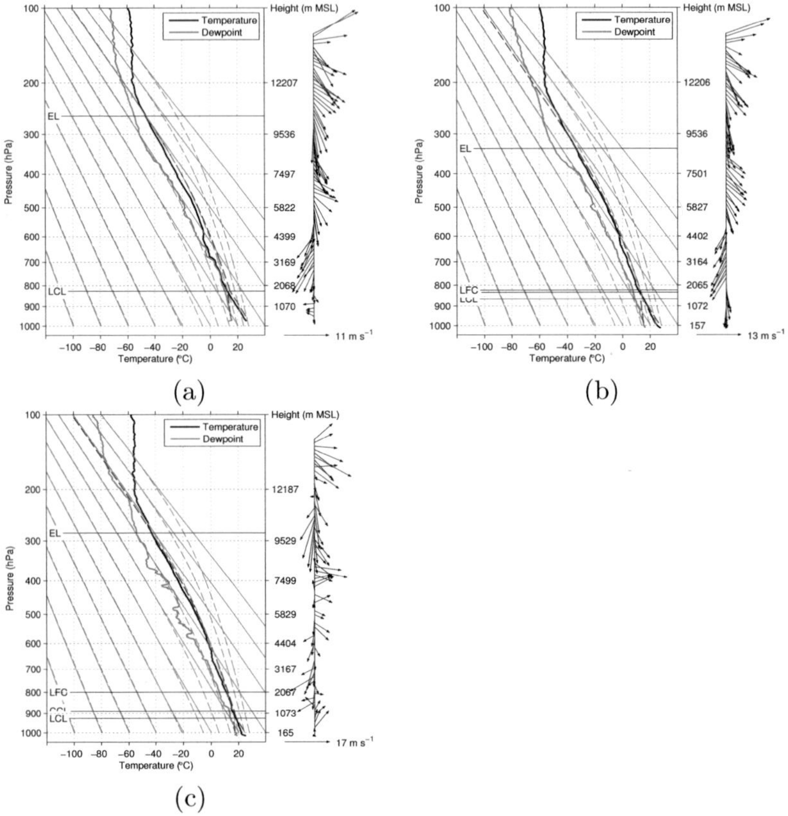

Figure B.11.: Radiosonde profiles of temperature, dewpoint and wind vector at 1300 LT at Corte (a), at 1300 LT at San Giuliano (b) and at 1300 LT at Ajaccio (c) on 06 September. The dashed blue lines indicate the virtual temperature of an air parcel with the mean layer temperature and humidity values of the lowest 50 hPa, which is lifted from the near the surface to its Lifting Condensation Level (LCL), Level of Free Convection (LFC) and Equilibrium Level (EL).

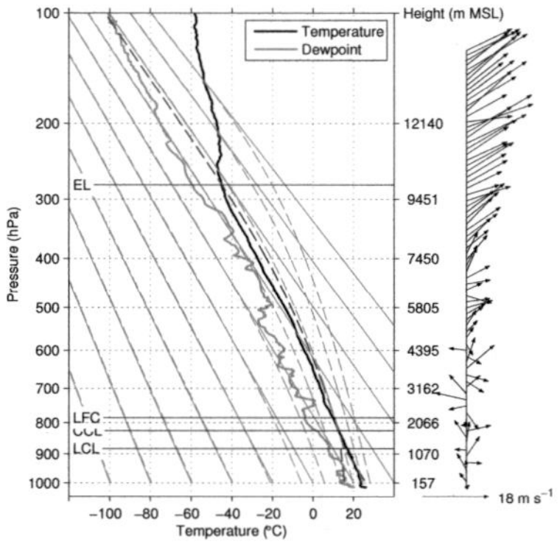

Figure B.12.: Radiosonde profiles of temperature, dewpoint and wind vector at 1300 LT at Ajaccio on 09 September. The dashed blue lines indicate the virtual temperature of an air parcel with the mean layer temperature and humidity values of the lowest 50 hPa, which is lifted from the near the surface to its Lifting Condensation Level (LCL), Level of Free Convection (LFC) and Equilibrium Level (EL).

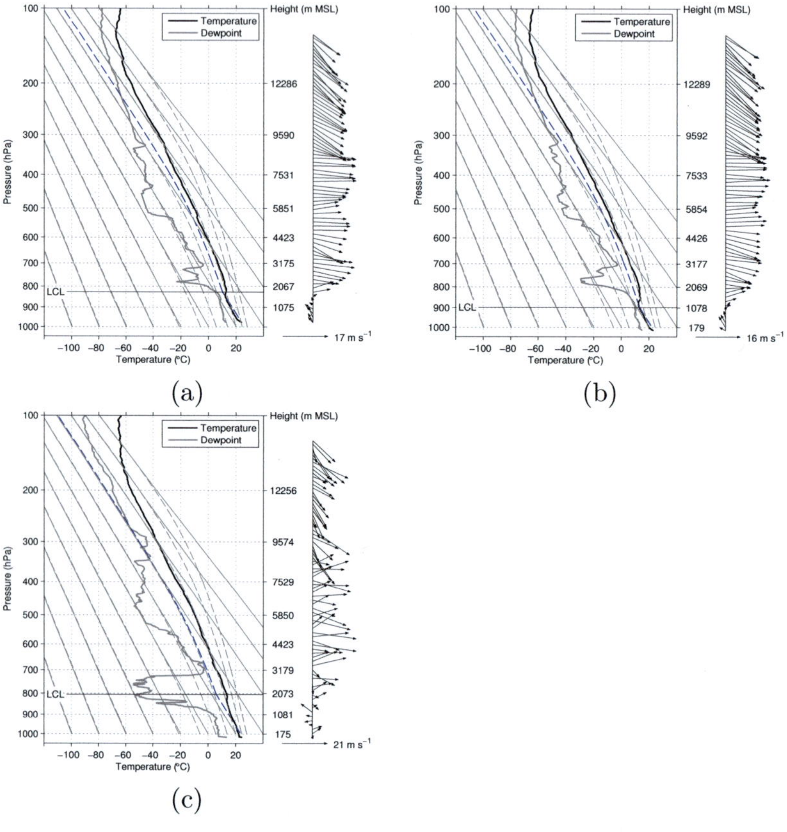

Figure B.13.: Radiosonde profiles of temperature, dewpoint and wind vector at 1500 LT at Corte (a), at 1500 LT at San Giuliano (b) and at 1300 LT at Ajaccio (c) on 21 September. The dashed blue lines indicate the virtual temperature of an air parcel with the mean layer temperature and humidity values of the lowest 50 hPa, which is lifted from the near the surface to its Lifting Condensation Level (LCL), Level of Free Convection (LFC) and Equilibrium Level (EL).

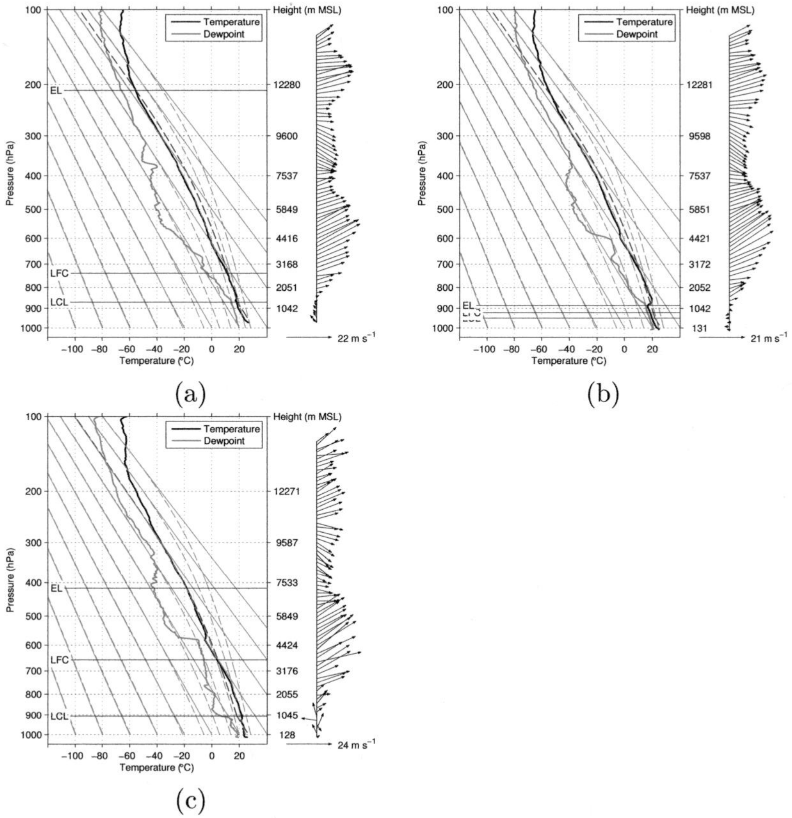

Figure B.14.: Radiosonde profiles of temperature, dewpoint and wind vector at 1500 LT at Corte (a), at 1500 LT at San Giuliano (b) and at 1300 LT at Ajaccio (c) on 23 September. The dashed blue lines indicate the virtual temperature of an air parcel with the mean layer temperature and humidity values of the lowest 50 hPa, which is lifted from the near the surface to its Lifting Condensation Level (LCL), Level of Free Convection (LFC) and Equilibrium Level (EL).

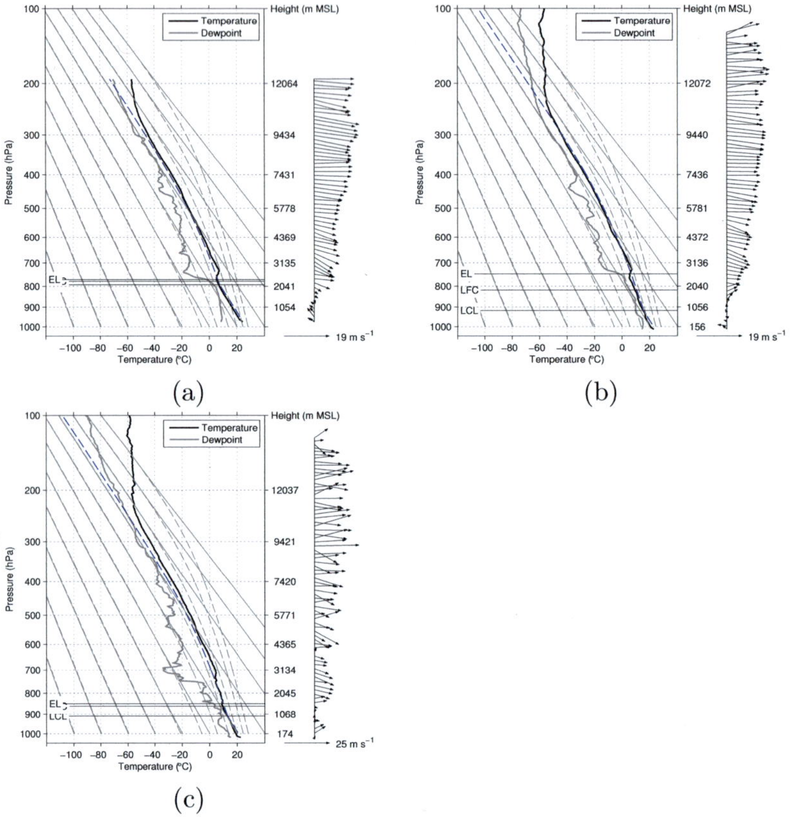

Figure B.15.: Radiosonde profiles of temperature, dewpoint and wind vector at 1500 LT at Corte (a), at 1500 LT at San Giuliano (b) and at 1300 LT at Ajaccio (c) on 02 October. The dashed blue lines indicate the virtual temperature of an air parcel with the mean layer temperature and humidity values of the lowest 50 hPa, which is lifted from the near the surface to its Lifting Condensation Level (LCL), Level of Free Convection (LFC) and Equilibrium Level (EL).

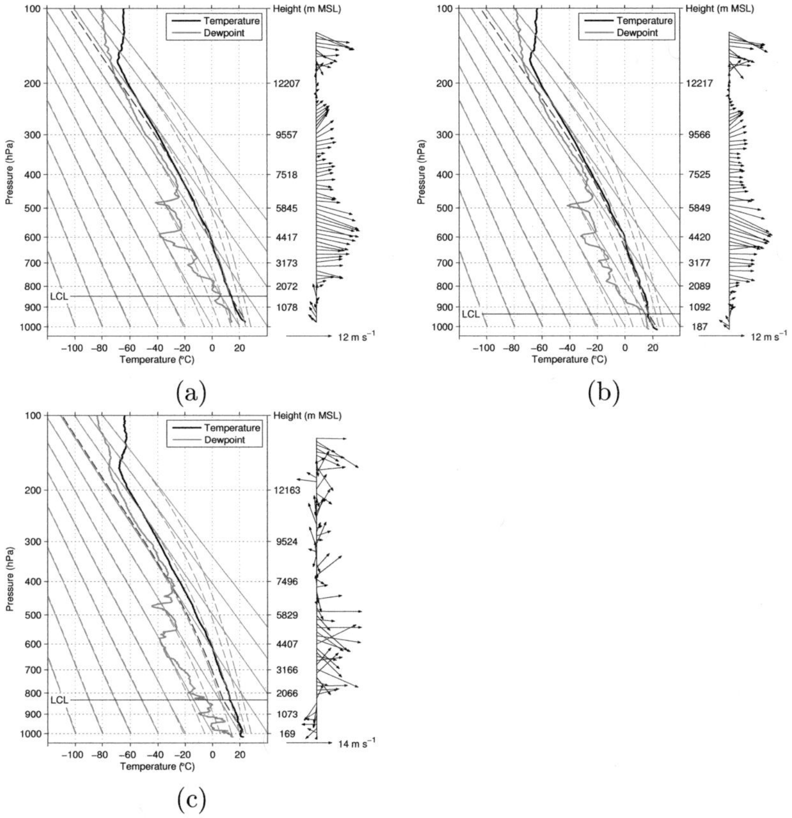

Figure B.16.: Radiosonde profiles of temperature, dewpoint and wind vector at 1500 LT at Corte (a), at 1500 LT at San Giuliano (b) and at 1300 LT at Ajaccio (c) on 05 October. The dashed blue lines indicate the virtual temperature of an air parcel with the mean layer temperature and humidity values of the lowest 50 hPa, which is lifted from the near the surface to its Lifting Condensation Level (LCL), Level of Free Convection (LFC) and Equilibrium Level (EL).

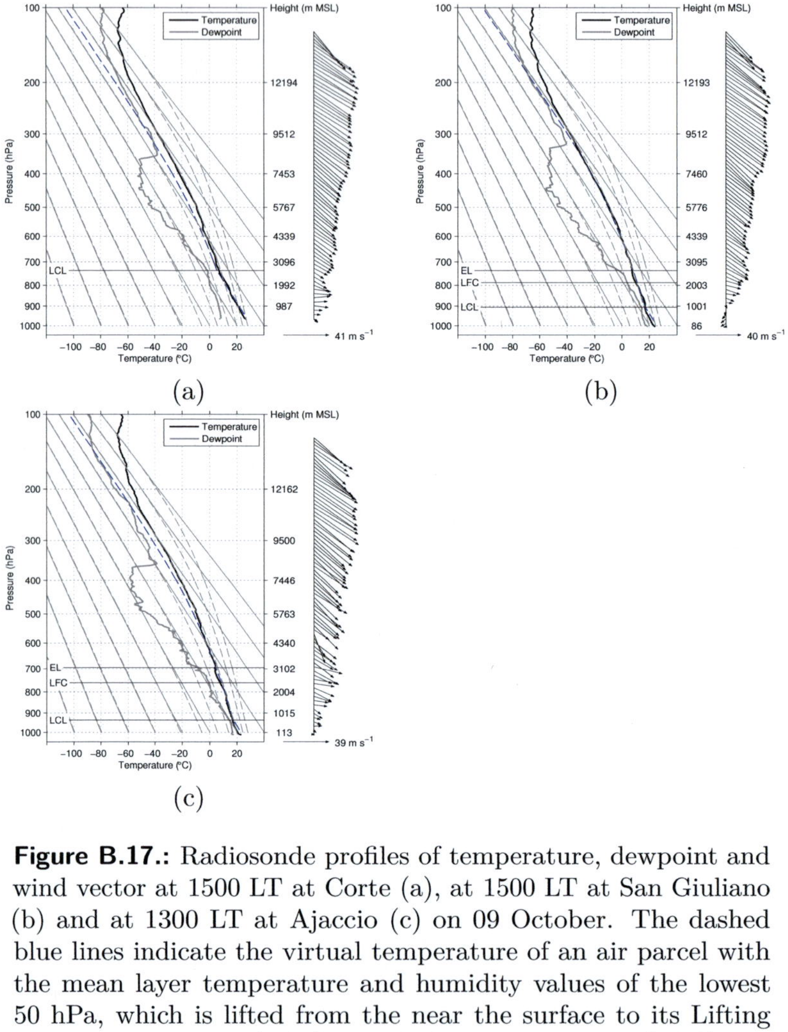

Figure B.17.: Radiosonde profiles of temperature, dewpoint and wind vector at 1500 LT at Corte (a), at 1500 LT at San Giuliano (b) and at 1300 LT at Ajaccio (c) on 09 October. The dashed blue lines indicate the virtual temperature of an air parcel with the mean layer temperature and humidity values of the lowest 50 hPa, which is lifted from the near the surface to its Lifting Condensation Level (LCL), Level of Free Convection (LFC) and Equilibrium Level (EL).

C. Bibliography

Adler, B., N. Kalthoff, and L. Gantner, 2011a: The impact of soil moisture inhomogeneities on the modification of a mesoscale convective system: an idealised model study. *Atmos. Res.*, **101 (1)**, 354–372.

———, 2011b: Initiation of deep convection caused by land-surface inhomogeneities in West Africa: a modelled case study. *Meteorol. Atmos. Phys.*, **112 (1-2)**, 15–27.

Arritt, R. W., J. M. Wilczak, and G. S. Young, 1992: Observations and numerical modeling of an elevated mixed layer. *Mon. Weather Rev.*, **120 (12)**, 2869–2880.

Atkinson, B. W., 1981: *Meso-scale atmospheric circulations.* Academic Press, London, 495 pp.

Banta, R. M., 1990: The role of mountain flows in making clouds. *Atmospheric processes over complex terrain*, Blumen, W., Ed., Boston, Meteorol. Monographs, 229–282.

Banta, R. M., L. D. Olivier, and D. H. Levinson, 1993: Evolution of the Monterey Bay sea-breeze layer as observed by pulsed doppler lidar. *J. Atmos. Sci.*, **50 (24)**, 3959–3982.

Banta, R. M., C. M. Shun, D. C. Law, W. Brown, R. F. Reinking, R. M. Hardesty, C. J. Senff, W. A. Brewer, M. J. Post, and L. S. Darby, 2013: Observational techniques: sampling the mountain atmosphere. *Mountain weather research and forecasting*, Chow, F. K., S. F. J. De Wekker, and B. J. Snyder, Eds., Springer, Dordrecht, Springer atmospheric sciences, 409–530.

Barthlott, C., U. Corsmeier, C. Meißner, F. Braun, and C. Kottmeier, 2006: The influence of mesoscale circulation systems on triggering convective cells over complex terrain. *Atmos. Res.*, **81 (2)**, 150–175.

Barthlott, C., J. W. Schipper, N. Kalthoff, B. Adler, C. Kottmeier, A. Blyth, and S. Mobbs, 2010: Model representation of boundary-layer convergence triggering deep convection over complex terrain: a case study from COPS. *Atmos. Res.*, **95 (2-3)**, 172–185.

Bastin, S., C. Champollion, O. Bock, P. Drobinski, and F. Masson, 2007: Diurnal cycle of water vapor as documented by a dense GPS network in a coastal area during ESCOMPTE IOP2. *J. Appl. Meteorol. Climatol.*, **46 (2)**, 167–182.

Bergström, H. and N. Juuso, 2006: A study of valley winds using the MIUU meso-scale model. *Wind Energy*, **9 (1-2)**, 109–129.

Bevis, M., S. Businger, S. Chiswell, T. A. Herring, R. A. Anthes, C. Rocken, and R. H. Ware, 1994: GPS meteorology: mapping zenith wet delays onto precipitable water. *J. Appl. Meteorol.*, **33 (3)**, 379–386.

Bevis, M., S. Businger, T. A. Herring, C. Rocken, R. A. Anthes, and R. H. Ware, 1992: GPS meteorology: remote sensing of atmospheric water vapor using the Global Positioning System. *J. Geophys. Res.*, **97 (D14)**, 15 787–15 801.

Beyrich, F., 1994: *Bestimmung der Mischungsschichthöhe aus Sodar-Daten unter Verwendung numerischer Modellrechnungen*, Schriftenreihe des Fraunhofer-Instituts für atmosphärische Umweltforschung, Vol. 28. Wiss.-Verlag Maraun, Frankfurt/Main, zugl.: Berlin, Freie Univ., Diss., 1994.

———, 1997: Mixing height estimation from sodar data - a critical discussion. *Atmos. Environ.*, **31 (23)**, 3941–3953.

Beyrich, F. and J.-P. Leps, 2012: An operational mixing height data set from routine radiosoundings at Lindenberg: Methodology. *Meteorol. Z.*, **21 (4)**, 337–348.

Bischoff-Gauß, I., N. Kalthoff, and M. Fiebig-Wittmaack, 2006: The influence of a storage lake in the arid Elqui Valley in Chile on local climate. *Theor. Appl. Climatol.*, **85 (3-4)**, 227–241.

Bischoff-Gauß, I., N. Kalthoff, and F. Fiedler, 1998: The impact of secondary flow systems on air pollution in the area of Sao Paulo. *J. Appl. Meteorol.*, **37 (3)**, 269–287.

Bischoff-Gauß, I., N. Kalthoff, S. Khodayar, M. Fiebig-Wittmaack, and S. Montecinos, 2008: Model simulations of the boundary-layer evolution over an arid Andes valley. *Boundary-Layer Meteorol.*, **128 (3)**, 357–379.

Bock, O., M.-N. Bouin, A. Walpersdorf, J. Lafore, S. Janicot, F. Guichard, and A. Agusti-Panareda, 2007: Comparison of ground-based GPS precipitable water vapour to independent observations and Numerical Weather Prediction model reanalyses over Africa. *Q. J. R. Meteorol. Soc.*, **133**, 2011–2027.

Bock, O., C. Keil, E. Richard, C. Flamant, and M. N. Bouin, 2005: Validation of precipitable water from ECMWF model analyses with GPS and radiosonde data during the MAP SOP. *Q. J. R. Meteorol. Soc.*, **131 (612)**, 3013–3036.

Bolton, D., 1980: The computation of equivalent potential temperature. *Mon. Weather Rev.*, **108 (7)**, 1046–1053.

Bougeault, P., P. Binder, A. Buzzi, R. Dirks, J. Kuettner, R. Houze, R. Smith, R. Steinacker, and H. Volkert, 2001: The MAP special observing period. *Bull. Amer. Meteorol. Soc.*, **82 (3)**, 433–462.

Brinkmann, W. A. R., 1971: What is a foehn? *Weather*, **26 (6)**, 230–240.

Brown, R. G. and C. Zhang, 1997: Variability of midtropospheric moisture and its effect on cloud-top height distribution during TOGA COARE*. *J. Atmos. Sci.*, **54 (23)**, 2760–2774.

Browning, K. A., A. M. Blyth, P. A. Clark, U. Corsmeier, C. J. Morcrette, J. L. Agnew, S. P. Ballard, D. Bamber, C. Barthlott, L. J. Bennett, et al., 2007: The convective storm initiation project. *Bull. Amer. Meteorol. Soc.*, **88 (12)**, 1939–1955.

Buettner, K. J. and N. Thyer, 1965: Valley winds in the Mount Rainier area. *Archiv für Meteorologie, Geophysik und Bioklimatologie, Serie B*, **14 (2)**, 125–147.

Buzzi, A. and L. Foschini, 2000: Mesoscale meteorological features associated with heavy precipitation in the southern Alpine region. *Meteorol. Atmos. Phys.*, **72 (2-4)**, 131–146.

Carlson, T. and F. Ludlam, 1968: Conditions for the occurrence of severe local storms. *Tellus*, **20 (2)**, 203–226.

Chaboureau, J.-P., F. Guichard, J.-L. Redelsperger, and J.-P. Lafore, 2004: The role of stability and moisture in the diurnal cycle of convection over land. *Q. J. R. Meteorol. Soc.*, **130 (604)**, 3105–3117.

Charba, J. P., 1979: Two to six hour severe local storm probabilities: an operational forecasting system. *Mon. Weather Rev.*, **107 (3)**, 268–282.

Chen, Y., C. Zhao, Q. Zhang, Z. Deng, M. Huang, and X. Ma, 2009: Aircraft study of mountain chimney effect of Beijing, China. *J. Geophys. Res.*, **114 (D8)**.

Ching, J., S. Shipley, and E. Browell, 1988: Evidence for cloud venting of mixed layer ozone and aerosols. *Atmos. Environ.*, **22 (2)**, 225–242.

Clarke, R., 1955: Some observations and comments on the sea breeze. *Aust. Meteorol. Mag.*, **11**, 47–68.

Corsmeier, U., R. Hankers, and A. Wieser, 2001: Airborne turbulence measurements in the lower troposphere onboard the research aircraft Dornier 128-6, D-IBUF. *Meteorol. Z.*, **10 (4)**, 315–329.

Corsmeier, U., N. Kalthoff, O. Kolle, M. Kotzian, and F. Fiedler, 1997: Ozone concentration jump in the stable nocturnal boundary layer during a LLJ-event. *Atmos. Environ.*, **31 (13)**, 1977–1989.

Cox, J. A., 2006: The sensitivity of thermally driven mountain flows to land cover change. Ph.D. thesis, The University of Utah, Salt Lake City.

Crewell, S. and U. Lohnert, 2007: Accuracy of boundary layer temperature profiles retrieved with multifrequency multiangle microwave radiometry. *Geoscience and Remote Sensing, IEEE Transactions on*, **45 (7)**, 2195–2201.

Crewell, S., U. Löhnert, M. Mech, and C. Simmer, 2011: Mikrowellenradiometrie für Wasserdampf- und Wolkenbeobachtungen. *Promet*, **36 (3/4)**, 109–118.

Crook, N. A., 1996: Sensitivity of moist convection forced by boundary layer processes to low-level thermodynamic fields. *Mon. Weather Rev.*, **124 (8)**, 1767–1785.

Crosman, E. T. and J. D. Horel, 2010: Sea and lake breezes: a review of numerical studies. *Boundary-Layer Meteorol.*, **137 (1)**, 1–29.

Dalla Fontana, A., 2008: Tuning of a thunderstorm index for north-eastern Italy. *Meteor. Appl.*, **15 (4)**, 475–482.

Davidson, B., 1963: Some turbulence and wind variability observations in the lee of mountain ridges. *J. Appl. Meteorol.*, **2 (4)**, 463–472.

De Wekker, S. F. J., 2002: Structure and morphology of the convective boundary layer in mountainous terrain. Ph.D. thesis, University of British Columbia.

———, 2008: Observational and numerical evidence of depressed convective boundary layer heights near a mountain base. *J. Appl. Meteorol. Climatol.*, **47**, 1017–1026.

De Wekker, S. F. J., D. Steyn, and S. Nyeki, 2004: A comparison of aerosol-layer and convective boundary-layer structure over a mountain range during STAAARTE'97. *Boundary-Layer Meteorol.*, **113 (2)**, 249–271.

Defant, F., 1949: Zur Theorie der Hangwinde, nebst Bemerkungen zur Theorie der Berg-und Talwinde. *Archiv für Meteorologie, Geophysik und Bioklimatologie, Serie A*, **1 (3-4)**, 421–450.

Derbyshire, S., I. Beau, P. Bechtold, J.-Y. Grandpeix, J.-M. Piriou, J.-L. Redelsperger, and P. Soares, 2004: Sensitivity of moist convection to environmental humidity. *Q. J. R. Meteorol. Soc.*, **130 (604)**, 3055–3079.

Doswell III, C. A., C. Ramis, R. Romero, and S. Alonso, 1998: A diagnostic study of three heavy precipitation episodes in the western Mediterranean region. *Wea. Forecasting*, **13 (1)**, 102–124.

Drobinski, P., V. Ducrocq, P. Alpert, E. Anagnostou, K. Béranger, M. Borga, I. Braud, A. Chanzy, S. Davolio, G. Delrieu, C. Estournel, N. Filali Boubrahmi, J. Font, V. Grubisic, S. Gualdi, V. Homar, B. Ivancan-Picek, C. Kottmeier, V. Kotroni, K. Lagouvardos, P. Lionello, M. Llasat, W. Ludwig, C. Lutoff, A. Mariotti, E. Richard, R. Romero, R. Rotunno, O. Roussot, I. Ruin, S. Somot, I. Taupier-Letage, J. Tintore, R. Uijlenhoet, and W. H., 2014: HyMeX, a 10-year multidisciplinary program on the Mediterranean water cycle. *Bull. Amer. Meteorol. Soc.*, **95**, 1063–1082.

Ducrocq, V., I. Braud, S. Davolio, R. Ferretti, C. Flamant, A. Jansa, N. Kalthoff, E. Richard, I. Taupier-Letage, P.-A. Ayral, S. Belamari, A. Berne, M. Borga, B. Boudevillain, O. Bock, J.-L. Boichard, M.-N. Bouin, O. Bousquet, C. Bouvier, J. Chiggiato, D. Cimini, U. Corsmeier, L. Coppola, P. Cocquerez, E. Defer, J. Delanoë, P. Di Girolamo, A. Doerenbecher, P. Drobinski, Y. Dufournet, N. Fourrié, J. J. Gourley, L. Labatut, D. Lambert, J. Le Coz, F. S. Marzano, G. Molinié, A. Montani, G. Nord, M. Nuret, K. Ramage, B. Rison, O. Roussot, F. Saïd, A. Schwarzenboeck, P. Testor, J. van Baelen, B. Vincendon, M. Aran, and J. Tamayo, 2014: HyMeX-SOP1, the field campaign dedicated to heavy precipitation and flash flooding in the northwestern Mediterranean. *Bull. Amer. Meteorol. Soc.*, **95**, 1083–1100.

Ducrocq, V., O. Nuissier, D. Ricard, C. Lebeaupin, and T. Thouvenin, 2008: A numerical study of three catastrophic precipitating events over southern France. II: mesoscale triggering and stationarity factors. *Q. J. R. Meteorol. Soc.*, **134 (630)**, 131–145.

Ducrocq, V., D. Tzanos, and S. Sénési, 1998: Diagnostic tools using a mesoscale NWP model for the early warning of convection. *Meteor. Appl.*, **5 (4)**, 329–349.

Durran, D., 2003: Downslope winds. *Encyclopedia of atmospheric sciences*, Holton, J. R., J. Curry, and P. J., Eds., Academic Press, Oxford, 644–650.

Durran, D. R., 1990: Mountain waves and downslope winds. *Atmospheric processes over complex terrain*, Blumen, W., Ed., Amer. Meteorol. Soc., Boston, Meteorol. Monographs, Vol. 23, 59–81.

Emanuel, K. A., 1994: *Atmospheric convection.* Oxford University Press, New York, 580 pp.

Emeis, S., K. Schafer, and C. Munkel, 2008: Surface-based remote sensing of the mixing-layer height - a review. *Meteorol. Z.*, **17 (5)**, 621–630.

Fast, J. D. and S. Zhong, 1998: Meteorological factors associated with inhomogeneous ozone concentrations within the Mexico City basin. *J. Geophys. Res.*, **103 (D15)**, 18 927–18 946.

Fiedler, F., I. Bischoff-Gauß, N. Kalthoff, and G. Adrian, 2000: Modeling of the transport and diffusion of a tracer in the Freiburg-Schauinsland area. *J. Geophys. Res.*, **105 (D1)**, 1599–1610.

Finkele, K., 1998: Inland and offshore propagation speeds of a sea breeze from simulations and measurements. *Boundary-Layer Meteorol.*, **87 (2)**, 307–329.

Freytag, C., 1987: Results from the MERKUR experiment: Mass budget and vertical motions in a large valley during mountain and valley wind. *Meteorol. Atmos. Phys.*, **37 (2)**, 129–140.

Frioud, M., V. Mitev, R. Matthey, H. Richner, M. Furger, and S. Gubser, 2004: Variation of the aerosol stratification over the Rhine Valley during foehn development: a backscatter lidar study. *Meteorol. Z.*, **13 (3)**, 175–181.

Furger, M., J. Dommen, W. K. Graber, L. Poggio, A. S. Prévôt, S. Emeis, G. Grell, T. Trickl, B. Gomiscek, B. Neininger, et al., 2000: The VOTALP Mesolcina Valley Campaign 1996 - concept, background and some highlights. *Atmos. Environ.*, **34 (9)**, 1395–1412.

Garratt, J. R., 1994: *The atmospheric boundary layer.* Cambridge atmospheric and space science series, Cambridge University Press, Cambridge, 316 pp.

Gerbier, N. and M. Berenger, 1961: Experimental studies of lee waves in the French Alps. *Q. J. R. Meteorol. Soc.*, **87 (371)**, 13–23.

Gohm, A., F. Harnisch, J. Vergeiner, F. Obleitner, R. Schnitzhofer, A. Hansel, A. Fix, B. Neininger, S. Emeis, and K. Schäfer, 2009: Air pollution transport in an Alpine valley: results from airborne and ground-based observations. *Boundary-Layer Meteorol.*, **131 (3)**, 441–463.

Gohm, A., G. Zängl, and G. J. Mayr, 2004: South foehn in the Wipp Valley on 24 October 1999 (MAP IOP 10): verification of high-resolution numerical simulations with observations. *Mon. Weather Rev.*, **132 (1)**, 78–102.

Graziano, T. M. and T. N. Carlson, 1987: A statistical evaluation of lid strength on deep convection. *Wea. Forecasting*, **2 (2)**, 127–139.

Grenzhäuser, J., 2011: *Entwicklung neuartiger Mess-und Auswertungsstrategien für ein scannendes Wolkenradar und deren Anwendungsbereiche*, Wissenschaftliche Berichte des Instituts für Meteorologie und Klimaforschung des Karlsruher Instituts für Technologie, Vol. 55. KIT Scientific Publishing, Karlsruhe, 119 pp., zugl.: Karlsruhe, KIT, Diss., 2010.

Grubišic, V., J. D. Doyle, J. Kuettner, R. Dirks, S. A. Cohn, L. L. Pan, S. Mobbs, R. B. Smith, C. D. Whiteman, S. Czyzyk, et al., 2008: The Terrain-Induced Rotor Experiment: a field campaign overview including observational highlights. *Bull. Amer. Meteorol. Soc.*, **89 (10)**, 1513–1533.

Gubser, S. and H. Richner, 2001: Investigations into mechanisms leading to the removal of the cold pool in foehn situations. *Extended abstracts from MAP meeting at Schliersee, Germany. MAP Newsletter 15.*, available at: http://www.map.meteoswiss.ch/map-doc/NL15/gubser2.pdf.

Hagemann, S., L. Bengtsson, and G. Gendt, 2003: On the determination of atmospheric water vapor from GPS measurements. *J. Geophys. Res.*, **108 (D21)**.

Haiden, T., 2003: On the pressure field in the slope wind layer. *J. Atmos. Sci.*, **60 (13)**, 1632–1635.

Handwerker, J., K. Träumner, J. Grenzhäuser, and A. Wieser, 2008: Simultaneous wind measurements with lidar and cloud radar - complementarity and quality check. *Proceedings of the Fifth European Conference on Radar in Meteorology and Hydrology*, Helsinki, Finnland, 26–30.

Henne, S., M. Furger, S. Nyeki, M. Steinbacher, B. Neininger, S. De Wekker, J. Dommen, N. Spichtinger, A. Stohl, and A. Prévôt, 2004: Quantification of topographic venting of boundary layer air to the free troposphere. *Atmos. Chem. Phys.*, **4 (2)**, 497–509.

Henne, S., M. Furger, and A. S. Prévôt, 2005: Climatology of mountain venting–induced elevated moisture layers in the lee of the Alps. *J. Appl. Meteorol.*, **44 (5)**, 620–633.

Hennemuth, B. and A. Lammert, 2006: Determination of the atmospheric boundary layer height from radiosonde and lidar backscatter. *Boundary-Layer Meteorol.*, **120**, 181–200.

Hogan, R. J., A. L. M. Grant, A. J. Illingworth, and G. N. Pearson, 2009: Vertical velocity variance and skewness in clear and cloud-topped boundary layers as revealed by doppler lidar. *Q. J. R. Meteorol. Soc.*, **135**, 635–643.

Homar, V., C. Ramis, R. Romero, S. Alonso, J. García-Moya, and M. Alarcón, 1999: A case of convection development over the western Mediterranean Sea: a study through numerical simulations. *Meteorol. Atmos. Phys.*, **71 (3-4)**, 169–188.

Iwasaki, H. and T. Miki, 2001: Observational study on the diurnal variation in precipitable water associated with the thermally induced local circulation over the semi-basin around maebashi using GPS data. *J. Meteor. Soc. Japan*, **79 (5)**, 1077–1091.

Jackson, P. L., G. Mayr, and S. Vosper, 2013: Dynamically-driven winds. *Mountain weather research and forecasting*, Chow, F. K., S. F. De Wekker, and B. J. Snyder, Eds., Springer, Dordrecht, Springer atmospheric sciences, 121–218.

Janssen, M. A., 1993: *Atmospheric remote sensing by microwave radiometry.* Wiley Series in Remote Sensing, John Wiley and Sons, New York, 572 pp.

Jiang, Q. and J. D. Doyle, 2008: Diurnal variation of downslope winds in Owens Valley during the Sierra Rotor Experiment. *Mon. Weather Rev.*, **136 (10)**, 3760–3780.

Jiang, Q., J. D. Doyle, and R. B. Smith, 2005: Blocking, descent and gravity waves: Observations and modelling of a MAP northerly föhn event. *Q. J. R. Meteorol. Soc.*, **131 (606)**, 675–701.

———, 2006: Interaction between trapped waves and boundary layers. *J. Atmos. Sci.*, **63**, 617–633.

Jiang, Q., R. B. Smith, and J. D. Doyle, 2008: Impact of the atmospheric boundary layer on mountain waves. *J. Atmos. Sci.*, **65 (2)**, 592–608.

Kalthoff, N., B. Adler, C. Barthlott, U. Corsmeier, S. Mobbs, S. Crewell, K. Träumner, C. Kottmeier, A. Wieser, V. Smith, and P. Di Girolamo, 2009: The impact of convergence zones on the initiation of deep convection: a case study from COPS. *Atmos. Res.*, **93 (4)**, 680–694.

Kalthoff, N., B. Adler, A. Wieser, M. Kohler, K. Träumner, J. Handwerker, U. Corsmeier, S. Khodayar, D. Lambert, A. Kopmann, et al., 2013a: KITcube – a mobile observation platform for convection studies deployed during HyMeX. *Meteorol. Z.*, **22 (6)**, 633–647.

Kalthoff, N., H.-J. Binder, M. Kossmann, R. Vögtlin, U. Corsmeier, F. Fiedler, and H. Schlager, 1998: Temporal evolution and spatial variation of the boundary layer over complex terrain. *Atmos. Environ.*, **32 (7)**, 1179–1194.

Kalthoff, N., M. Fiebig-Wittmaack, C. Meißner, M. Kohler, M. Uriarte, I. Bischoff-Gauß, and E. Gonzales, 2006: The energy balance, evapo-transpiration and nocturnal dew deposition of an arid valley in the Andes. *J. Arid Environ.*, **65 (3)**, 420–443.

Kalthoff, N., V. Horlacher, U. Corsmeier, A. Volz-Thomas, B. Kolahgar, H. Geiß, M. Möllmann-Coers, and A. Knaps, 2000: Influence of valley winds on transport and dispersion of airborne pollutants in the Freiburg-Schauinsland area. *J. Geophys. Res.*, **105 (D1)**, 1585–1597.

Kalthoff, N., M. Kohler, C. Barthlott, B. Adler, S. Mobbs, U. Corsmeier, K. Träumner, T. Foken, R. Eigenmann, L. Krauss, S. Khodayar, and P. Di Girolamo, 2011: The dependence of convection-related parameters on surface and boundary-layer conditions over complex terrain. *Q. J. R. Meteorol. Soc.*, **137 (S1)**, 70–80.

Kalthoff, N., K. Träumner, B. Adler, S. Späth, A. Behrendt, A. Wieser, J. Handwerker, F. Madonna, and V. Wulfmeyer, 2013b: Dry and moist convection in the boundary layer over the Black Forest - a combined analysis of in situ and remote sensing data. *Meteorol. Z.*, **22**, 445–461.

Keulegan, G., 1957: An experimental study of the motion of saline water from locks into fresh water channels. *Natl. Bur. Stand. Rep.*, **5168**.

Khodayar, S., 2009: High-resolution analysis of the initiation of deep convection forced by boundary-layer processes. Ph.D. thesis, Universität Karlsruhe, Karlsruhe.

Khodayar, S., N. Kalthoff, M. Fiebig-Wittmaack, and M. Kohler, 2008: Evolution of the atmospheric boundary-layer structure of an arid Andes valley. *Meteorol. Atmos. Phys.*, **99 (3-4)**, 181–198.

Klemp, J. and D. Lilly, 1975: The dynamics of wave-induced downslope winds. *J. Atmos. Sci.*, **32 (2)**, 320–339.

Kneifel, S., 2008: Modellierung und Beobachtung von horizontalen Wasserdampfinhomogenitäten mit Hilfe von Mikrowellenradiometrie. Master's thesis, Universität zu Köln, Köln.

Kondo, H., 1990a: A numerical experiment of the extended sea breeze over the Kanto plain. *J. Meteor. Soc. Japan*, **68 (4)**, 419–434.

———, 1990b: A numerical experiment on the interaction between sea breeze and valley wind to generate the so-called extended sea breeze. *J. Meteor. Soc. Japan*, **68 (4)**, 435–446.

Kondo, J., T. Kuwagata, and S. Haginoya, 1989: Heat budget analysis of nocturnal cooling and daytime heating in a basin. *J. Atmos. Sci.*, **46 (19)**, 2917–2933.

Kossmann, M., U. Corsmeier, S. F. J. De Wekker, F. Fiedler, R. Vögtlin, N. Kalthoff, H. Güsten, and B. Neininger, 1999: Observations of handover processes between the atmospheric boundary layer and the free troposphere over mountainous terrain. *Contr. Atmos. Phys.*, **72**, 329–350.

Kottmeier, C., P. Palacio-Sese, N. Kalthoff, U. Corsmeier, and F. Fiedler, 2000: Sea breezes and coastal jets in southeastern Spain. *Int. J. Climatol.*, **20 (14)**, 1791–1808.

Kraus, H., 2004: *Die Atmosphäre der Erde: Eine Einführung in die Meteorologie*. Springer Berlin, Heidelberg, 422 pp.

Kuang, Z. and C. S. Bretherton, 2006: A mass-flux scheme view of a high-resolution simulation of a transition from shallow to deep cumulus convection. *J. Atmos. Sci.*, **63 (7)**, 1895–1909.

Kunz, M., 2007: The skill of convective parameters and indices to predict isolated and severe thunderstorms. *Nat. Hazards Earth Syst. Sci.*, **7 (2)**, 327–342.

Kuwagata, T., A. Numaguti, and N. Endo, 2001: Diurnal variation of water vapor over the central Tibetan Plateau during summer. *J. Meteor. Soc. Japan*, **79 (1B)**, 401–418.

Lambert, D., M. Mallet, V. Ducrocq, F. Dulac, F. Gheusi, and N. Kalthoff, 2011: CORSiCA: a Mediterranean atmospheric and oceanographic observatory in Corsica within the framework of HyMeX and ChArMEx. *Adv. Geosci.*, **26**, 125–131.

Lascaux, F., E. Richard, C. Keil, and O. Bock, 2004: Impact of the MAP reanalysis on the numerical simulation of the MAP-IOP2a convective system. *Meteorol. Z.*, **13 (1)**, 49–54.

LeMone, M. A., 1990: Some observations of vertical velocity skewness in the convective planetary boundary layer. *J. Atmos. Sci.*, **47**, 1163—1169.

Lenschow, D. H., J. Mann, and L. Kristensen, 1994: How long is long enough when measuring fluxes and other turbulence statistics? *J. Atmos. Oceanic. Technol.*, **11**, 661–673.

Lenschow, D. H., V. Wulfmeyer, and C. Senff, 2000: Measuring second- through fourth-order moments in noisy data. *J. Atmos. Oceanic. Technol.*, **17**, 1330–1347.

Lenschow, D. H., J. C. Wyngaard, and W. T. Pennell, 1980: Mean-field and second-moment budgets in a baroclinic, convective boundary layer. *J. Atmos. Sci.*, **37**, 1313–1326.

Löhnert, U. and S. Crewell, 2003: Accuracy of cloud liquid water path from ground-based microwave radiometry 1. dependency on cloud model statistics. *Radio Science*, **38 (3)**.

Löhnert, U., D. Turner, and S. Crewell, 2009: Ground-based temperature and humidity profiling using spectral infrared and microwave observations. Part I: simulated retrieval performance in clear-sky conditions. *J. Appl. Meteorol. Climatol.*, **48 (5)**, 1017–1032.

Long, R. R., 1954: Some aspects of the flow of stratified fluids: II. experiments with a two-fluid system. *Tellus*, **6 (2)**, 97–115.

Lu, R. and R. P. Turco, 1994: Air pollutant transport in a coastal environment. Part I: two-dimensional simulations of sea-breeze and mountain effects. *J. Atmos. Sci.*, **51 (15)**, 2285–2308.

Markowski, P. and Y. Richardson, 2010: *Mesoscale meteorology in midlatitudes*. John Wiley & Sons, Ltd, Chichester, UK, 407 pp.

Mauder, M. and T. Foken, 2011: *Documentation and instruction manual of the eddy-covariance software package TK3*, Vol. 46. Arbeitsergebnisse Universität Bayreuth, Abt. Mikrometeorologie, Universität of Bayreuth, Bayreuth, 58 pp.

Mayr, G. J. and L. Armi, 2010: The influence of downstream diurnal heating on the descent of flow across the Sierras. *J. Appl. Meteorol. Climatol.*, **49**, 1906–1912.

McGowan, H. A., 2004: Observations of anti-winds in a deep alpine valley, Lake Tekapo, New Zealand. *Arctic, Antarctic, and Alpine Res*, **36 (4)**, 495–501.

Meißner, C., N. Kalthoff, M. Kunz, and G. Adrian, 2007: Initiation of shallow convection in the Black Forest mountains. *Atmos. Res.*, **86 (1)**, 42–60.

Metzger, J., C. Barthlott, and N. Kalthoff, 2014: Impact of upstream flow conditions on the initiation of moist convection over the island of Corsica. *Atmos. Res.*, **145–146**, 279 – 296.

Millán, M. M., M. José Sanz, R. Salvador, and E. Mantilla, 2002: Atmospheric dynamics and ozone cycles related to nitrogen deposition in the western Mediterranean. *Environ. Pollution*, **118 (2)**, 167–186.

Miller, R. C., 1972: Notes on analysis and severe-storm forecasting procedures of the Air Force Global Weather Central. Tech. Rep. 200, Air Wea. Service.

Miller, S., B. Keim, R. Talbot, and H. Mao, 2003: Sea breeze: structure, forecasting, and impacts. *Rev. Geophys.*, **41 (3)**.

Nappo, C. J., 2013: *An introduction to atmospheric gravity waves.* 2d ed., Academic Press, Amsterdam, 359 pp.

Nater, W., H. Richner, and P. D. Phillips, 1979: Shear instabiiities and their characteristics during foehn. *Geophys. Astrophys. Fluid Dyn.*, **13 (1)**, 215–223.

Nyeki, S., K. Eleftheriadis, U. Baltensperger, I. Colbeck, M. Fiebig, A. Fix, C. Kiemle, M. Lazaridis, and A. Petzold, 2002: Airborne lidar and in-situ aerosol observations of an elevated layer, leeward of the European Alps and Apennines. *Geophys. Res. Lett.*, **29 (17)**, 33–1–33–4.

Ohtani, R., 2001: Detection of water vapor variations driven by thermally-induced local circulations using the Japanese Continuous GPS Array. *Geophys. Res. Lett.*, **28 (1)**, 151–154.

Orlanski, I., 1975: A rational subdivision of scales for atmospheric processes. *Bull. Amer. Meteorol. Soc.*, **56**, 527–530.

Peltier, W. and T. Clark, 1979: The evolution and stability of finite-amplitude mountain waves. Part II: surface wave drag and severe downslope windstorms. *J. Atmos. Sci.*, **36 (8)**, 1498–1529.

Physick, W. L. and D. J. Abbs, 1992: Flow and plume dispersion in a coastal valley. *J. Appl. Meteorol.*, **31 (1)**, 64–73.

Pospichal, B. and S. Crewell, 2007: Boundary layer observations in West Africa using a novel microwave radiometer. *Meteorol. Z.*, **16 (5)**, 513–523.

Poulos, G. S., J. E. Bossert, T. B. McKee, and R. A. Pielke, 2000: The interaction of katabatic flow and mountain waves. Part I: observations and idealized simulations. *J. Atmos. Sci.*, **57 (12)**, 1919–1936.

Poulos, G. S., J. E. Bossert, T. B. McKee, and R. A. Pielke Sr, 2007: The interaction of katabatic flow and mountain waves. Part II: case study analysis and conceptual model. *J. Atmos. Sci.*, **64 (6)**, 1857–1879.

Prévôt, A. S., J. Dommen, and M. Bäumle, 2000: Influence of road traffic on volatile organic compound concentrations in and above a deep Alpine valley. *Atmos. Environ.*, **34 (27)**, 4719–4726.

Qian, J.-H., 2008: Why precipitation is mostly concentrated over islands in the maritime continent. *J. Atmos. Sci.*, **65 (4)**, 1428–1441.

Rampanelli, G. and D. Zardi, 2004: A method to determine the capping inversion of the convective boundary layer. *J. Appl. Meteorol.*, **43 (6)**, 925–933.

Rampanelli, G., D. Zardi, and R. Rotunno, 2004: Mechanisms of up-valley winds. *J. Atmos. Sci.*, **61**, 3097–3111.

Redelsperger, J.-L., D. Parsons, and F. Guichard, 2002: Recovery processes and factors limiting cloud-top height following the arrival of a dry intrusion observed during TOGA COARE. *J. Atmos. Sci.*, **59 (16)**, 2438–2457.

Reiter, R., H. Müller, R. Sladkovic, and K. Munzert, 1984: Aerologische Untersuchungen des tagesperiodischen Windsystems im Inntal während MERKUR. *Meteorol. Rundschau*, **37 (5)**, 176–190.

Reuten, C., D. Steyn, K. Strawbridge, and P. Bovis, 2005: Observations of the relation between upslope flows and the convective boundary layer in steep terrain. *Boundary-Layer Meteorol.*, **116 (1)**, 37–61.

Ricard, D., V. Ducrocq, and L. Auger, 2012: A climatology of the mesoscale environment associated with heavily precipitating events over a northwestern Mediterranean area. *J. Appl. Meteorol. Climatol.*, **51 (3)**, 468–488.

Richard, E., P. Mascart, and E. C. Nickerson, 1989: The role of surface friction in downslope windstorms. *J. Appl. Meteorol.*, **28 (4)**, 241–251.

Richner, H. and P. Hächler, 2013: Understanding and forecasting alpine foehn. *Mountain weather research and forecasting*, Chow, F. K., S. F. De Wekker, and B. J. Snyder, Eds., Springer, Dordrecht, Springer atmospheric sciences, 219–260.

Rocken, C., T. V. Hove, J. Johnson, F. Solheim, R. Ware, M. Bevis, S. Chiswell, and S. Businger, 1995: GPS/STORM-GPS sensing of atmospheric water vapor for meteorology. *J. Atmos. Oceanic. Technol.*, **12 (3)**, 468–478.

Rocken, C., R. Ware, T. Van Hove, F. Solheim, C. Alber, J. Johnson, M. Bevis, and S. Businger, 1993: Sensing atmospheric water vapor with the Global Positioning System. *Geophys. Res. Lett.*, **20 (23)**, 2631–2634.

Rose, T., S. Crewell, U. Löhnert, and C. Simmer, 2005: A network suitable microwave radiometer for operational monitoring of the cloudy atmosphere. *Atmos. Res.*, **75 (3)**, 183–200.

Rotach, M. W., M. Andretta, P. Calanca, A. P. Weigel, and A. Weiss, 2008: Boundary layer characteristics and turbulent exchange mechanisms in highly complex terrain. *Acta Geophys.*, **56**, 194–219.

Rotach, M. W. and D. Zardi, 2007: On the boundary-layer structure over highly complex terrain: Key findings from MAP. *Q. J. R. Meteorol. Soc.*, **133 (625)**, 937–948.

Rotunno, R. and R. Ferretti, 2001: Mechanisms of intense Alpine rainfall. *J. Atmos. Sci.*, **58 (13)**, 1732–1749.

Saastamoinen, J., 1972: Atmospheric correction for the troposphere and stratosphere in radio ranging satellites. *The use of artificial satellites for geodesy*, Henriksen, S. W., A. Mancini, and B. H. Chovitz, Eds., American Geophysical Union, Washington, D. C., 247–251.

Sasaki, T., P. Wu, S. Mori, J.-I. Hamada, Y. I. Tauhid, M. D. Yamanaka, T. Sribimawati, T. Yoshikane, and F. Kimura, 2004:

Vertical moisture transport above the mixed layer around the mountains in western Sumatra. *Geophys. Res. Lett.*, **31** (**8**).

Seibert, P., F. Beyrich, S.-E. Gryning, S. Joffre, A. Rasmussen, and P. Tercier, 2000: Review and intercomparison of operational methods for the determination of the mixing height. *Atmos. Environ.*, **34** (**7**), 1001–1027.

Serafin, S. and D. Zardi, 2010a: Daytime heat transfer processes related to slope flows and turbulent convection in an idealized mountain valley. *J. Atmos. Sci.*, **67**, 3739–3756.

———, 2010b: Structure of the atmospheric boundary layer in the vicinity of a developing upslope flow system: a numerical model study. *J. Atmos. Sci.*, **67** (**4**), 1171–1185.

———, 2011: Daytime development of the boundary layer over a plain and in a valley under fair weather conditions: a comparison by means of idealized numerical simulations. *J. Atmos. Sci.*, **68** (**9**), 2128–2141.

Simpson, J., 1969: A comparison between laboratory and atmospheric density currents. *Q. J. R. Meteorol. Soc.*, **95** (**406**), 758–765.

Simpson, J. and R. Britter, 1980: A laboratory model of an atmospheric mesofront. *Q. J. R. Meteorol. Soc.*, **106** (**449**), 485–500.

Simpson, J., D. Mansfield, and J. Milford, 1977: Inland penetration of sea-breeze fronts. *Q. J. R. Meteorol. Soc.*, **103** (**435**), 47–76.

Smith, R. B., 1980: Linear theory of stratified hydrostatic flow past an isolated mountain. *Tellus*, **32** (**4**), 348–364.

———, 2007: Interacting mountain waves and boundary layers. *J. Atmos. Sci.*, **64** (**2**), 594–607.

Smith, R. B., Q. Jiang, and J. D. Doyle, 2006: A theory of gravity wave absorption by a boundary layer. *J. Atmos. Sci.*, **63**, 774–781.

Smith, R. B., S. Skubis, J. D. Doyle, A. S. Broad, C. Kiemle, and H. Volkert, 2002: Mountain waves over Mont Blanc: influence of a stagnant boundary layer. *J. Atmos. Sci.*, **59**, 2073–2092.

Solheim, F. S., J. Vivekanandan, R. H. Ware, and C. Rocken, 1999: Propagation delays induced in GPS signals by dry air, water vapor, hydrometeors, and other particulates. *J. Geophys. Res.*, **104 (D8)**, 9663–9670.

Späth, S., 2012: Thermische Windsysteme im westlichen Mittelmeerraum. Master's thesis, Karlsruher Institut für Technologie, Karlsruhe.

Steinacker, R., 1984: Area-height distribution of a valley and its relation to the valley wind. *Contr. Atmos. Phys.*, **57 (1)**, 64–71.

Steyn, D. G., M. Baldi, and R. Hoff, 1999: The detection of mixed layer depth and entrainment zone thickness from lidar backscatter profiles. *J. Atmos. Oceanic. Technol.*, **16 (7)**, 953–959.

Stull, R. B., 1988: *An introduction to boundary layer meteorology.* Kluwer Academic Publishers, Dordrecht, 666 pp.

———, 2000: *Meteorology for scientists and engineers.* 2d ed., Brooks/Cole Thomson Learning, Pacific Grove, Calif., 502 pp.

Sun, J. and S. F. J. De Wekker, 2011: Atmospheric carbon dioxide transport over mountainous terrain. *Mountain ecosystems: Dynamics, management and conservation*, Richards, K. E., Ed., Nova Science Publishers, Inc., New York, chap. 5, 101–121.

Takagi, T., F. Kimura, and S. Kono, 2000: Diurnal variation of GPS precipitable water at Lhasa in premonsoon and monsoon periods. *J. Meteor. Soc. Japan*, **78 (2)**, 175–180.

Träumner, K., 2012: *Einmischprozesse am Oberrand der konvektiven atmosphärischen Grenzschicht*, Wissenschaftliche Berichte des Instituts für Meteorologie und Klimaforschung des Karlsruher Instituts für Technologie, Vol. 51. KIT Scientific Publishing, Karlsruhe, 245 pp., zugl.: Karlsruhe, KIT, Diss., 2010.

Träumner, K., C. Kottmeier, U. Corsmeier, and A. Wieser, 2011: Convective boundary-layer entrainment: short review and progress using doppler lidar. *Boundary-Layer Meteorol.*, **141 (3)**, 369–391.

Tucker, S. C., C. J. Senff, A. M. Weickmann, W. A. Brewer, R. M. Banta, S. P. Sandberg, D. C. Law, and R. M. Hardesty, 2009: Doppler lidar estimation of mixing height using turbulence, shear, and aerosol profiles. *J. Atmos. Oceanic. Technol.*, **26 (4)**, 673–688.

Vergeiner, I. and E. Dreiseitl, 1987: Valley winds and slope winds - observations and elementary thoughts. *Meteorol. Atmos. Phys.*, **36 (1-4)**, 264–286.

Vogt, S. and G. Jaubert, 2004: Foehn in the Rhine Valley as seen by a wind-profiler-RASS system and comparison with the nonhydrostatic model Meso-NH. *Meteorol. Z.*, **13 (3)**, 165–174.

Vosper, S. and A. Brown, 2007: The effect of small-scale hills on orographic drag. *Q. J. R. Meteorol. Soc.*, **133 (627)**, 1345–1352.

Wagner, A., 1932: Neue Theorie des Berg- und Talwindes. *Meteorol. Z.*, **49 (9)**, 329–341.

Weckwerth, T. M., 2000: The effect of small-scale moisture variability on thunderstorm initiation. *Mon. Weather Rev.*, **128 (12)**, 4017–4030.

Weckwerth, T. M., D. B. Parsons, S. E. Koch, J. A. Moore, M. A. LeMone, B. B. Demoz, C. Flamant, B. Geerts, J. Wang, and W. F. Feltz, 2004: An overview of the International H[sub 2] O Project (IHOP_2002) and some preliminary highlights. *Bull. Amer. Meteorol. Soc.*, **85 (2)**, 253–277.

Weckwerth, T. M., J. W. Wilson, and R. M. Wakimoto, 1996: Thermodynamic variability within the convective boundary layer due to horizontal convective rolls. *Mon. Weather Rev.*, **124 (5)**, 769–784.

Weigel, A. P., F. K. Chow, and M. W. Rotach, 2007a: The effect of mountainous topography on moisture exchange between the "surface" and the free atmosphere. *Boundary-Layer Meteorol.*, **125 (2)**, 227–244.

———, 2007b: On the nature of turbulent kinetic energy in a steep and narrow Alpine valley. *Boundary-Layer Meteorol.*, **123 (1)**, 177–199.

Weigel, A. P., F. K. Chow, M. W. Rotach, R. L. Street, and M. Xue, 2006: High-resolution large-eddy simulations of flow in a steep Alpine valley. Part II: flow structure and heat budgets. *J. Appl. Meteorol. Climatol.*, **45**, 87–107.

Weigel, A. P. and M. W. Rotach, 2004: Flow structure and turbulence characteristics of the daytime atmosphere in a steep and narrow Alpine valley. *Q. J. R. Meteorol. Soc.*, **130 (602)**, 2605–2627.

Whiteman, C. D., 1982: Breakup of temperature inversions in deep mountain valleys. *J. Appl. Meteorol.*, **21**, 270–289.

———, 2000: *Mountain meteorology fundamentals and applications.* Oxford University Press, New York, 355 pp.

Whiteman, C. D. and J. C. Doran, 1993: The relationship between overlying synoptic-scale flows and winds within a valley. *J. Appl. Meteorol.*, **32 (11)**, 1669–1682.

Whiteman, C. D. and T. B. McKee, 1982: Breakup of temperature inversions in deep mountain valleys: Part II. Thermodynamic model. *J. Appl. Meteorol.*, **21 (3)**, 290–302.

Wickert, J. and G. Gendt, 2006: Fernerkundung der Erdatmosphäre mit GPS. *Promet*, **32 (3/4)**, 176–184.

Wooldridge, G. L. and E. L. McIntyre, 1986: The dynamics of the planetary boundary layer over a heated mountain slope. *Geofizika*, **3 (1)**, 3–21.

Wu, P., J.-I. Hamada, S. Mori, Y. I. Tauhid, M. D. Yamanaka, and F. Kimura, 2003: Diurnal variation of precipitable water over a mountainous area of Sumatra Island. *J. Appl. Meteorol.*, **42 (8)**, 1107–1115.

Zängl, G., A. Gohm, and G. Geier, 2004: South foehn in the Wipp Valley - Innsbruck region: numerical simulations of the 24 October 1999 case (MAP-IOP 10). *Meteorol. Atmos. Phys.*, **86 (3-4)**, 213–243.

Zardi, D. and C. Whiteman, 2013: Diurnal mountain wind systems. *Mountain weather research and forecasting*, Chow, F. K., S. F. J. De Wekker, and B. J. Snyder, Eds., Springer, Dordrecht, Springer atmospheric sciences, 35–119.

Wissenschaftliche Berichte des Instituts für Meteorologie und Klimaforschung des Karlsruher Instituts für Technologie (0179-5619)

Bisher erschienen:

Nr. 1: *Fiedler, F. / Prenosil, T.*
Das MESOKLIP-Experiment. (Mesoskaliges Klimaprogramm im Oberrheintal).
August 1980

Nr. 2: *Tangermann-Dlugi, G.*
Numerische Simulationen atmosphärischer Grenzschichtströmungen über langgestreckten mesoskaligen Hügelketten bei neutraler thermischer Schichtung.
August 1982

Nr. 3: *Witte, N.*
Ein numerisches Modell des Wärmehaushalts fließender Gewässer unter Berücksichtigung thermischer Eingriffe.
Dezember 1982

Nr. 4: *Fiedler, F. / Höschele, K. (Hrsg.)*
Prof. Dr. Max Diem zum 70. Geburtstag.
Februar 1983 (vergriffen)

Nr. 5: *Adrian, G.*
Ein Initialisierungsverfahren für numerische mesoskalige Strömungsmodelle.
Juli 1985

Nr. 6: *Dorwarth, G.*
Numerische Berechnung des Druckwiderstandes typischer Geländeformen.
Januar 1986

Nr. 7: *Vogel, B.; Adrian, G. / Fiedler, F.*
MESOKLIP-Analysen der meteorologischen Beobachtungen von mesoskaligen Phänomenen im Oberrheingraben.
November 1987

Nr. 8: *Hugelmann, C.-P.*
Differenzenverfahren zur Behandlung der Advektion.
Februar 1988

Nr. 9: *Hafner, T.*
Experimentelle Untersuchung zum Druckwiderstand der Alpen.
April 1988

Nr. 10: *Corsmeier, U.*
Analyse turbulenter Bewegungsvorgänge in der maritimen atmosphärischen Grenzschicht.
Mai 1988

Nr. 11: *Walk, O. / Wieringa, J.(eds)*
Tsumeb Studies of the Tropical Boundary-Layer Climate.
Juli 1988

Nr. 12: *Degrazia, G. A.*
Anwendung von Ähnlichkeitsverfahren auf die turbulente Diffusion in der konvektiven und stabilen Grenzschicht.
Januar 1989

Nr. 13: *Schädler, G.*
Numerische Simulationen zur Wechselwirkung zwischen Landoberflächen und atmophärischer Grenzschicht.
November 1990

Nr. 14: *Heldt, K.*
Untersuchungen zur Überströmung eines mikroskaligen Hindernisses in der Atmosphäre.
Juli 1991

Nr. 15: *Vogel, H.*
Verteilungen reaktiver Luftbeimengungen im Lee einer Stadt – Numerische Untersuchungen der relevanten Prozesse.
Juli 1991

Nr. 16: *Höschele, K.(ed.)*
Planning Applications of Urban and Building Climatology – Proceedings of the IFHP / CIB-Symposium Berlin, October 14-15, 1991.
März 1992

Nr. 17: *Frank, H. P.*
Grenzschichtstruktur in Fronten.
März 1992

Nr. 18: *Müller, A.*
Parallelisierung numerischer Verfahren zur Beschreibung von Ausbreitungs- und chemischen Umwandlungsprozessen in der atmosphärischen Grenzschicht.
Februar 1996

Nr. 19: *Lenz, C.-J.*
Energieumsetzungen an der Erdoberfläche in gegliedertem Gelände.
Juni 1996

Nr. 20: *Schwartz, A.*
Numerische Simulationen zur Massenbilanz chemisch reaktiver Substanzen im mesoskaligen Bereich.
November 1996

Nr. 21: *Beheng, K. D.*
Professor Dr. Franz Fiedler zum 60. Geburtstag.
Januar 1998

Nr. 22: *Niemann, V.*
Numerische Simulation turbulenter Scherströmungen mit einem Kaskadenmodell.
April 1998

Nr. 23: *Koßmann, M.*
Einfluß orographisch induzierter Transportprozesse auf die Struktur der atmosphärischen Grenzschicht und die Verteilung von Spurengasen.
April 1998

Nr. 24: *Baldauf, M.*
Die effektive Rauhigkeit über komplexem Gelände – Ein Störungstheoretischer Ansatz.
Juni 1998

Nr. 25: *Noppel, H.*
Untersuchung des vertikalen Wärmetransports durch die Hangwindzirkulation auf regionaler Skala.
Dezember 1999

Nr. 26: *Kuntze, K.*
Vertikaler Austausch und chemische Umwandlung von Spurenstoffen über topographisch gegliedertem Gelände.
Oktober 2001

Nr. 27: *Wilms-Grabe, W.*
Vierdimensionale Datenassimilation als Methode zur Kopplung zweier verschiedenskaliger meteorologischer Modellsysteme.
Oktober 2001

Nr. 28: *Grabe, F.*
Simulation der Wechselwirkung zwischen Atmosphäre, Vegetation und Erdoberfläche bei Verwendung unterschiedlicher Parametrisierungsansätze.
Januar 2002

Nr. 29: *Riemer, N.*
Numerische Simulationen zur Wirkung des Aerosols auf die troposphärische Chemie und die Sichtweite.
Mai 2002

Nr. 30: *Braun, F. J.*
Mesoskalige Modellierung der Bodenhydrologie.
Dezember 2002

Nr. 31: *Kunz, M.*
Simulation von Starkniederschlägen mit langer Andauer über Mittelgebirgen.
März 2003

Nr. 32: *Bäumer, D.*
Transport und chemische Umwandlung von Luftschadstoffen im Nahbereich von Autobahnen – numerische Simulationen.
Juni 2003

Nr. 33: *Barthlott, C.*
Kohärente Wirbelstrukturen in der atmosphärischen Grenzschicht.
Juni 2003

Nr. 34: *Wieser, A.*
Messung turbulenter Spurengasflüsse vom Flugzeug aus.
Januar 2005

Nr. 35: *Blahak, U.*
Analyse des Extinktionseffektes bei Niederschlagsmessungen mit einem C-Band Radar anhand von Simulation und Messung.
Februar 2005

Nr. 36: *Bertram, I.*
Bestimmung der Wasser- und Eismasse hochreichender konvektiver Wolken anhand von Radardaten, Modellergebnissen und konzeptioneller Betrachtungen.
Mai 2005

Nr. 37: *Schmoeckel, J.*
Orographischer Einfluss auf die Strömung abgeleitet aus Sturmschäden im Schwarzwald während des Orkans „Lothar".
Mai 2006

Nr. 38: *Schmitt, C.*
Interannual Variability in Antarctic Sea Ice Motion: Interannuelle Variabilität antarktischer Meereis-Drift.
Mai 2006

Nr. 39: *Hasel, M.*
Strukturmerkmale und Modelldarstellung der Konvektion über Mittelgebirgen.
Juli 2006

Ab Band 40 erscheinen die Wissenschaftlichen Berichte des Instituts für Meteorologie und Klimaforschung bei KIT Scientific Publishing (ISSN 0179-5619).
Die Bände sind unter www.ksp.kit.edu als PDF frei verfügbar oder als Druckausgabe bestellbar.

Nr. 40: *Lux, R.*
Modellsimulationen zur Strömungsverstärkung von orographischen Grundstrukturen bei Sturmsituationen. (2007)
ISBN 978-3-86644-140-8

Nr. 41: *Straub, W.*
Der Einfluss von Gebirgswellen auf die Initiierung und Entwicklung konvektiver Wolken. (2008)
ISBN 978-3-86644-226-9

Nr. 42: *Meißner, C.*
High-resolution sensitivity studies with the regional climate model COSMO-CLM. (2008)
ISBN 978-3-86644-228-3

Nr. 43: *Höpfner, M.*
Charakterisierung polarer stratosphärischer Wolken mittels hochauflösender Infrarotspektroskopie. (2008)
ISBN 978-3-86644-294-8

Nr. 44: *Rings, J.*
Monitoring the water content evolution of dikes. (2009)
ISBN 978-3-86644-321-1

Nr. 45: *Riemer, M.*
Außertropische Umwandlung tropischer Wirbelstürme: Einfluss auf das Strömungsmuster in den mittleren Breiten. (2012)
ISBN 978-3-86644-766-0

Nr. 46: *Anwender, D.*
Extratropical Transition in the Ensemble Prediction System of the ECMWF: Case Studies and Experiments. (2012)
ISBN 978-3-86644-767-7

Nr. 47: *Rinke, R.*
Parametrisierung des Auswaschens von Aerosolpartikeln durch Niederschlag. (2012)
ISBN 978-3-86644-768-4

Nr. 48: *Stanelle, T.*
Wechselwirkungen von Mineralstaubpartikeln mit thermodynamischen und dynamischen Prozessen in der Atmosphäre über Westafrika. (2012)
ISBN 978-3-86644-769-1

Nr. 49: *Peters, T.*
Ableitung einer Beziehung zwischen der Radarreflektivität, der Niederschlagsrate und weiteren aus Radardaten abgeleiteten Parametern unter Verwendung von Methoden der multivariaten Statistik. (2012)
ISBN 978-3-86644-323-5

Nr. 50: *Khodayar Pardo, S.*
High-resolution analysis of the initiation of deep convection forced by boundary-layer processes. (2012)
ISBN 978-3-86644-770-7

Nr. 51: *Träumner, K.*
Einmischprozesse am Oberrand der konvektiven atmosphärischen Grenzschicht. (2012)
ISBN 978-3-86644-771-4

Nr. 52: *Schwendike, J.*
Convection in an African Easterly Wave over West Africa and the Eastern Atlantic: A Model Case Study of Hurricane Helene (2006) and its Interaction with the Saharan Air Layer. (2012)
ISBN 978-3-86644-772-1

Nr. 53: *Lundgren, K.*
Direct Radiative Effects of Sea Salt on the Regional Scale. (2012)
ISBN 978-3-86644-773-8

Nr. 54: *Sasse, R.*
Analyse des regionalen atmosphärischen Wasserhaushalts unter Verwendung von COSMO-Simulationen und GPS-Beobachtungen. (2012)
ISBN 978-3-86644-774-5

Nr. 55: *Grenzhäuser, J.*
Entwicklung neuartiger Mess- und Auswertungsstrategien für ein scannendes Wolkenradar und deren Anwendungsbereiche. (2012)
ISBN 978-3-86644-775-2

Nr. 56: *Grams, C.*
Quantification of the downstream impact of extratropical transition for Typhoon Jangmi and other case studies. (2013)
ISBN 978-3-86644-776-9

Nr. 57: *Keller, J.*
Diagnosing the Downstream Impact of Extratropical Transition Using Multimodel Operational Ensemble Prediction Systems. (2013)
ISBN 978-3-86644-984-8

Nr. 58: *Mohr, S.*
Änderung des Gewitter- und Hagelpotentials im Klimawandel. (2013)
ISBN 978-3-86644-994-7

Nr. 59: *Puskeiler, M.*
Radarbasierte Analyse der Hagelgefährdung in Deutschland. (2013)
ISBN 978-3-7315-0028-5

Nr. 60: *Zeng, Y.*
Efficient Radar Forward Operator for Operational Data Assimilation within the COSMO-model. (2014)
ISBN 978-3-7315-0128-2

Nr. 61: *Bangert, M. J.*
Interaction of Aerosol, Clouds, and Radiation on the Regional Scale. (2014)
ISBN 978-3-7315-0123-7

Nr. 62: *Jerger, D.*
Radar Forward Operator for Verification of Cloud Resolving Simulations within the COSMO Model. (2014)
ISBN 978-3-7315-0172-5

Nr. 63: *Maurer, V.*
Vorhersagbarkeit konvektiver Niederschläge: Hochauflösende Ensemblesimulationen für Westafrika. (2014)
ISBN 978-3-7315-0189-3

Nr. 64: *Stawiarski, C.*
Optimizing Dual-Doppler Lidar Measurements of Surface Layer Coherent Structures with Large-Eddy Simulations. (2014)
ISBN 978-3-7315-0197-8

Nr. 65: *Mahlke, H.*
Mechanismen der Auslösung hochreichender Konvektion im südwestdeutschen Mittelgebirgsraum. (2014)
ISBN 978-3-7315-0203-6

Nr. 66: *Fosser, G.*
Precipitation statistics from regional climate model at resolutions relevant for soil erosion. (2014)
ISBN 978-3-7315-0227-2

Nr. 67: *Adler, B.*
Boundary-Layer Processes Producing Mesoscale Water-Vapour Variability over a Mountainous Island. (2014)
ISBN 978-3-7315-0247-0